WORKING GUIDE TO PETROLEUM AND NATURAL GAS PRODUCTION ENGINEERING

WORKING GUIDE TO
PETROLEUM AND NATURAL GAS PRODUCTION ENGINEERING

—

WILLIAM C. LYONS

ELSEVIER

AMSTERDAM • BOSTON • HEIDELBERG • LONDON
NEW YORK • OXFORD • PARIS • SAN DIEGO
SAN FRANCISCO • SINGAPORE • SYDNEY • TOKYO

Gulf Publishing is an imprint of Elsevier

Gulf Publishing is an imprint of Elsevier
30 Corporate Drive, Suite 400, Burlington, MA 01803, USA
The Boulevard, Langford Lane, Oxford OX5 1GB

First edition 2010

Notice

No responsibility is assumed by the publisher for any injury and/or damage to persons or property as a matter of products liability, negligence or otherwise, or from any use or operation of any methods, products, instructions or ideas contained in the material herein. Because of rapid advances in the medical sciences, in particular, independent verification of diagnoses and drug dosages should be made.

Library of Congress Cataloging in Publication Data
A catalog record for this book is available from the Library of Congress

British Library Cataloguing in Publication Data
A catalogue record for this book is available from the British Library

ISBN: 978-1-85617-845-7

For information on all Elsevier publications
visit our website at *elsevierdirect.com*

Typeset by: diacriTech, India

Printed and bound in United States of America
10 11 12 13   11 10 9 8 7 6 5 4 3 2 1

Contents

Full Contents

Chapter 1

PROPERTIES OF HYDROCARBON MIXTURES

Chapter 2

FLOW OF FLUIDS

Chapter 3

NATURAL FLOW PERFORMANCE

Chapter 4

GAS PRODUCTION ENGINEERING

1

Properties of Hydrocarbon Mixtures

This chapter contains correlation and procedures for the prediction of physical properties of natural gas and oil. Physical constants of single components are given in Table 1.1.

1.1 COMPRESSIBILITY FACTOR AND PHASE BEHAVIOR

The compressibility factor Z is a dimensionless factor, independent of the quantity of gas and determined by the character of the gas, the temperature, and the pressure.

$$Z = \frac{PV}{nRT} = \frac{MPV}{mRT} \qquad (1.1)$$

TABLE 1.1 Physical Constants of Hydrocarbons

No.	Compound	Formula	1. Molecular Mass	2. Boiling point, °C 101.3250 kPa (abs)	Vapor pressure, kPa (abs) 40°C	3. Freezing point, °C 101.3250 kPa (abs)	Critical constants Pressure, kPa (abs)	Temperature, K	Volume, m³/kg
1	Methane	CH_4	16.043	−161.52(28)	(35,000.0)	−182.47[d]	4604.0	190.55	0.00617
2	Ethane	C_2H_6	30.070	−88.58	(6000.0)	−182.80[d]	4880.0	305.43	0.00492
3	Propane	C_3H_8	44.097	−42.07	1341.0	−187.68[d]	4249.0	369.82	0.00460
4	n-Butane	C_4H_{10}	58.124	−0.49	377.0	−138.36	3797.0	425.16	0.00439
5	Isobutane	C_4H_{10}	58.124	−11.81	528.0	−159.60	3648.0	408.13	0.00452
6	n-Pentane	C_5H_{12}	72.151	36.06	115.66	−129.73	3369.0	469.6	0.00421
7	Isopentane	C_5H_{12}	72.151	27.84	151.3	−159.90	3381.0	460.39	0.00424
8	Neopentane	C_5H_{12}	72.151	9.50	269.0	−16.55	3199.0	433.75	0.00420
9	n-Hexane	C_6H_{14}	86.178	68.74	37.28	−95.32	3012.0	507.4	0.00429
10	2-Methylpentane	C_6H_{14}	86.178	60.26	50.68	−153.66	3010.0	497.45	0.00426
11	3-Methylpentane	C_6H_{14}	86.178	63.27	45.73	—	3124.0	504.4	0.00426
12	Neohexane	C_6H_{14}	86.178	49.73	73.41	−99.870	3081.0	488.73	0.00417
13	2,3-Dimethylbutane	C_6H_{14}	86.178	57.98	55.34	−128.54	3127.0	499.93	0.00415
14	n-Heptane	C_7H_{16}	100.205	98.42	12.34	−90.582	2736.0	540.2	0.00431
15	2-Methylhexane	C_7H_{16}	100.205	90.05	17.22	−118.27	2734.0	530.31	0.00420
16	3-Methylhexane	C_7H_{16}	100.205	91.85	16.16	—	2814.0	535.19	0.00403
17	3-Ethylpentane	C_7H_{16}	100.205	93.48	15.27	−118.60	2891.0	540.57	0.00415
18	2,2-Dimethylpentane	C_7H_{16}	100.205	79.19	26.32	−123.81	2773.0	520.44	0.00415
19	2,4-Dimethylpentane	C_7H_{16}	100.205	80.49	24.84	−119.24	2737.0	519.73	0.00417
20	3,3-Dimethylpentane	C_7H_{16}	100.205	86.06	20.93	−134.46	2945.0	536.34	0.00413
21	Triptane	C_7H_{16}	100.205	80.88	25.40	−24.91	2954.0	531.11	0.00397
22	n-Octane	C_8H_{16}	114.232	125.67	4.143	−56.76	2486.0	568.76	0.00431
23	Diisobutyl	C_8H_{18}	114.232	109.11	8.417	−91.200	2486.0	549.99	0.00422

24	Isoctane	C_8H_{18}	114.232	99.24	12.96	-107.38	2568.0	543.89	0.00410
25	n-Nonane	C_9H_{20}	128.259	150.82	1.40	-53.49	2288.0	594.56	0.00427
26	n-Decane	$C_{10}H_{22}$	142.286	174.16	0.4732	-29.64	2099.0	617.4	0.00424
27	Cyclopentane	C_5H_{10}	70.135	49.25	73.97	-93.866	4502.0	511.6	0.00371
28	Methylcyclopentane	C_6H_{12}	84.162	71.81	33.85	-142.46	3785.0	532.73	0.00379
29	Cyclohexane	C_6H_{12}	84.162	80.73	24.63	6.554	4074.0	553.5	0.00368
30	Methylcyclohexane	C_7H_{14}	96.189	100.93	12.213	-126.59	3472.0	572.12	0.00375
31	Ethene (Ethylene)	C_2H_4	28.054	-103.77(29)	—	-169.15[d]	5041.0	282.35	0.00467
32	Propene (Propylene)	C_3H_6	42.081	-47.72	1596.0	-185.25[d]	4600.0	364.85	0.00430
33	1-Butene (Butylene)	C_4H_8	56.108	-6.23	451.9	-185.35[d]	4023.0	419.53	0.00428
34	cis-2-Butene	C_4H_8	56.108	3.72	337.6	-138.91	4220.0	435.58	0.00417
35	trans-2-Butene	C_4H_8	56.108	0.88	365.8	-105.55	4047.0	428.63	0.00424
36	Isobutene	C_4H_8	56.108	-6.91	452.3	-140.35	3999.0	417.90	0.00426
37	1-Pentene	C_5H_{10}	70.135	29.96	141.65	-165.22	3529.0	464.78	0.00422
38	1,2-Butadiene	C_4H_6	54.092	10.85	269.0	-136.19	(4502.0)	(444.0)	(0.00405)
39	1,3-Butadiene	C_4H_6	54.092	-4.41	434.0	-108.91	4330.0	425.0	0.00409
40	Isoprene	C_5H_8	68.119	34.07	123.77	-145.95	(3850.0)	(484.0)	(0.00406)
41	Acetylene	C_2H_2	26.038	-84.88[e]	—	-80.8[d]	6139.0	306.33	0.00434
42	Benzene	C_6H_6	78.114	80.09	24.38	5.533	4898.0	562.16	0.00328
43	Toluene	C_7H_8	92.141	110.63	7.895	-94.991	4106.0	591.80	0.00343
44	Ethylbenzene	C_8H_{10}	106.168	136.20	2.87	-94.975	3609.0	617.20	0.00353
45	o-Xylene	C_8H_{10}	106.168	144.43	2.05	-25.18	3734.0	630.33	0.00348
46	m-Xylene	C_8H_{10}	106.168	139.12	2.53	-47.87	3536.0	617.05	0.00354
47	p-Xylene	C_8H_{10}	106.168	138.36	2.65	13.26	3511.0	616.23	0.00356
48	Styrene	C_8H_8	104.152	145.14	1.85	-30.61	3999.0	647.6	0.00338
49	Isopropylbenzene	C_9H_{12}	120.195	152.41	1.47	-96.035	3209.0	631.1	0.00357
50	Methyl alcohol	CH_4O	32.042	64.54	35.43	-97.68	8096.0	512.64	0.00368
51	Ethyl alcohol	C_2H_6O	46.089	78.29	17.70	-114.1	6383.0	513.92	0.00362
52	Carbon monoxide	CO	28.010	-191.49	—	-205.0[d]	3499.0(33)	132.92(33)	0.00332(33)
53	Carbon dioxide	CO_2	44.010	-78.51[e]	—	-56.57[d]	7382.0(33)	304.19(33)	0.00214(33)

(Continued)

TABLE 1.1 (Continued)

No.	Compound	Formula	Molecular Mass	Boiling point, °C 101.3250 kPa (abs)	Vapor pressure, kPa (abs) 40°C	Freezing point, °C 101.3250 kPa (abs)	Critical constants Pressure, kPa (abs)	Critical constants Temperature, K	Critical constants Volume, m³/kg
	See Note No. → 2		1.	2.		3.			
54	Hydrogen sulfide	H_2S	34.076	−60.31	2881.0	−85.53[d]	9005.0	373.5	0.00287
55	Sulfur dioxide	SO_2	64.069	−10.02	630.8	−75.48[d]	7894.0	430.8	0.00190
56	Ammonia	NH_3	17.031	−33.33(30)	1513.0	−77.74[d]	11,280.0	405.6	0.00425
57	Air	$N_2 + O_2$	28.964	−194.2(2)	—	—	3771.0(2)	132.4(2)	0.00323(3)
58	Hydrogen	H_2	2.016	−252.87[v]	—	−259.2[d]	1297.0	33.2	0.03224
59	Oxygen	O_2	31.999	−182.962[v]	—	−218.8[d]	5061.0	154.7(33)	0.00229
60	Nitrogen	N_2	28.013	−195.80(31)	—	−210.0[d]	3399.0	126.1	0.00322
61	Chlorine	Cl_2	70.906	−34.03	1134.0	−101.0[d]	7711.0	417.0	0.00175
62	Water	H_2O	18.015	100.00[v]	7.377	0.00	22,118.0	647.3	0.00318
63	Helium	He	4.003	−268.93(32)	—	—	227.5(32)	5.2(32)	0.01436(32)
64	Hydrogen chloride	HCl	36.461	−85.00	6304.0	−114.18[d]	8309.0	324.7	0.00222

NOTES

a: Air saturated liquid.

b: Absolute values from weights in vacuum.

c: The apparent values from weight in air are shown for users' convenience and compliance with ASTM-IP Petroleum Measurement Tables. In the United States and Great Britain, all commercial weights are required by law to be weights in air. All other mass data are on an absolute mass (weight in vacuum) basis.

d: At saturation pressure (triple point).

e: Sublimation point.

f: The + sign and number following signify the ASTM octane number corresponding to that of 2,2,4-trimethylpentane with the indicated number of cm³ of TEL added per gal.

g: Determined at 100°C.

h: Saturation pressure and 15°C.

i: Apparent value at 15°C.

j: Average value from octane numbers of more than one sample.

k: Relative density (specific gravity), 48.3°C/15°C (sublimation point; solid C_2H_4/liquid H_2O).

m: Densities of liquid at the boiling point.

n: Heat of sublimation.

p: See Note 10.

s: Extrapolated to room temperature from higher temperature.

t: Gross calorific values shown for ideal gas volumes are not direct conversions of each other using only the gas volume per liquid volume value shown herein. The values differ by the heat of vaporization to ideal gas at 288.15 K.

v: Fixed points on the 1968 International Practical Temperature Scale (IPTS-68).

w: Value for normal hydrogen (25% para, 75% ortho). The value for equilibrium mixture of para and ortho is −0.218; however, in most correlations, 0 is used.

x: Densities at the boiling point in kg/m³ for: Ethane, 546.4; propane, 581.0; propene, 608.8; hydrogen sulfide, 960.0; sulfur dioxide, 1462.0; ammonia, 681.6; hydrogen chloride, 1192.

*: Calculated values.

(): Estimated values.

†: Values are estimated using 2nd virial coefficients.

TABLE 1.1 (Continued)

| Relative density 15°C/15°C[a,b] | Density of liquid 101.3250 kPa (abs), 15°C | | | Temperature coefficient of density, at 15°C, a −1°C | Pitzer acentric factor, a | Compressibility factor of real gas, Z 101.3250 kPa (abs), 15°C | Ideal gas* 101.3250 kPa (abs), 15°C | | | Specific heat capacity 101.3250 kPa (abs), 15°C c_p kJ/(kg·°C) | | No. |
	kg/m³,a (mass in vacuum)	kg/m³,a,c (Apparent mass in air)	m³/kmol				Relative density Air=1	Specific volume m³/kg	Volume ratio gas/(liquid in vacuum)	Ideal gas	Liquid	
(0.3)[j]	(300.0)[j]	(300.0)[j]	(0.05)[i]	—	0.0126	0.9981	0.5539	1.474	(442.0)[l]	2.204	—	1
0.3581[h]	357.8[h,x]	356.6[h]	0.084 04[h]	—	0.0978	0.9915	1.0382	0.7863	281.3[h]	1.706	3.807	2
0.5083[h]	507.8[h,x]	506.7[h]	0.086 84[h]	0.002 74[h]	0.1541	0.9810	1.5225	0.5362	272.3[h]	1.625	2.476	3
0.5847[h]	584.2[h]	583.1[h]	0.099 49[h]	0.002 11[h]	0.2015	0.9641	2.0068	0.4068	237.6[h]	1.652	2.366(41)	4
0.5637[h]	563.2[h]	562.1[h]	0.1032[h]	0.002 14[h]	0.1840	0.9665	2.0068	0.4068	229.1[h]	1.616	2.366(41)	5
0.6316	631.0	629.9	0.1143	0.00157	0.2524	0.942[+]	2.4911	0.3277	206.8	1.622	2.292(41)	6
0.6250	624.4	623.3	0.1156	0.001 62	0.2286	0.948[+]	2.4911	0.3277	204.6	1.600	2.239	7
0.5972[h]	596.7[h]	595.6[h]	0.1209[h]	0.001 87[h]	0.1967	0.9538	2.4911	0.3277	195.5[h]	1.624	2.317	8
0.6644	663.8	662.7	0.1298	0.00135	0.2998	0.910[+]	2.9753	0.2744	182.1	1.613	2.231	9
0.6583	657.7	656.6	0.1310	0.00140	0.2784	—	2.9753	0.2744	180.5	1.602	2.205	10
0.6694	668.8	667.7	0.1289	0.00135	0.2741	—	2.9753	0.2744	183.5	1.578	2.170	11
0.6545	653.9	652.8	0.1318	0.00140	0.2333	—	2.9753	0.2744	179.4	1.593	2.148	12
0.6668	666.2	665.1	0.1294	0.00135	0.2475	—	2.9753	0.2744	182.8	1.566	2.146	13
0.6886	688.0	686.9	0.1456	0.00124	0.3494	0.852[+]	3.4596	0.2360	162.4	1.606	2.209	14
0.6835	682.8	681.7	0.1468	0.00122	0.3303	—	3.4596	0.2360	161.1	1.595	2.183	15

(Continued)

TABLE 1.1 (Continued)

Relative density 15°C/15° C^a,b	Density of liquid 101.3250 kPa (abs), 15°C			Temperature coefficient of density, at 15°C, a −1/°C	Pitzer acentric factor, a	Compressibility factor of real gas, Z 101.3250 kPa (abs), 15°C	Ideal gas* 101.3250 kPa (abs), 15°C			Specific heat capacity 101.3250 kPa (abs), 15°C c_p kJ/(kg·°C)		No.
	kg/m³,a,c (mass in vacuum)	kg/m³,a,c (Apparent mass in air)	m³/kmol				Relative density Air=1	Specific volume m³/kg	Volume ratio gas/(liquid in vacuum)	Ideal gas	Liquid	
0.6921	691.5	690.4	0.1449	0.00124	0.3239	—	3.4596	0.2360	163.2	1.584	2.137	16
0.7032	702.6	701.5	0.1426	0.00126	0.3107	—	3.4596	0.2360	165.8	1.613	2.150	17
0.6787	678.0	676.9	0.1478	0.00130	0.2876	—	3.4596	0.2360	160.0	1.613	2.161	18
0.6777	677.1	676.0	0.1480	0.00130	0.3031	—	3.4596	0.2360	159.8	1.651	2.193	19
0.6980	697.4	696.3	0.1437	0.00117	0.2681	—	3.4596	0.2360	164.6	1.603	2.099	20
0.6950	694.4	693.3	0.1443	0.00124	0.2509	—	3.4596	0.2360	163.9	1.578	2.088	21
0.7073	706.7	705.6	0.1616	0.00112	0.3961	0.783†	3.9439	0.2070	146.3	1.601	2.191	22
0.6984	697.7	696.6	0.1637	0.00117	0.3564	—	3.9439	0.2070	144.4	1.573	2.138	23
0.6966	696.0	694.9	0.1641	0.00117	0.3041	—	3.9439	0.2070	144.1	1.599	2.049	24
0.7224	721.7	720.6	0.1777	0.00113	0.4452	—	4.4282	0.1843	133.0	1.598	2.184	25
0.7346	733.9	732.8	0.1939	0.00099	0.4904	—	4.9125	0.1662	122.0	1.595	2.179	26
0.7508	750.2	749.1	0.09349	0.00126	0.1945	0.949†	2.4215	0.3371	252.9	1.133	1.763	27
0.7541	753.4	752.3	0.1117	0.00128	0.2308	—	2.9057	0.2809	211.7	1.258	1.843	28
0.7838	783.1	782.0	0.1075	0.00122	0.2098	—	2.9057	0.2809	220.0	1.211	1.811	29
0.7744	773.7	772.6	0.1269	0.00113	0.2364	—	3.3900	0.2406	186.3	1.324	1.839	30
—	—	—	—	—	0.869	0.9938	0.9686	0.8428	—	1.514	—	31
0.5231^h	522.6^h,x	521.5^h	0.06069^h	0.00340^h	0.1443	0.9844	1.4529	0.5619	293.6^h	1.480	2.443	32
0.6019^h	601.4^h	600.3^h	0.09330^h	0.00209^h	0.1949	0.9703	1.9372	0.4214	253.4^h	1.483	2.237	33

0.6277^h	627.1^h	626.0^h	$0.089\,47^h$	$0.001\,76^h$	0.2033	0.9660	1.9372	0.4214	264.3^h	1.366	2.241(42)	34
0.6105^h	610.0^h	608.9^h	$0.091\,98^h$	$0.001\,93^h$	0.2126	0.9661	1.9372	0.4214	257.1^h	1.528	2.238	35
0.6010^h	600.5^h	599.4^h	$0.093\,44^h$	$0.002\,16^h$	0.2026	0.9688	1.9372	0.4214	253.1^h	1.547	2.296	36
0.6462	645.6	644.5	0.1086	0.001 60	0.2334	$0.948^†$	2.4215	0.3371	217.7	1.519	2.241(43)	37
0.6578^h	657.0^h	656.0^h	$0.082\,33^h$	$0.001\,76^h$	(0.2540)	(0.969)	1.8676	0.4371	287.2^h	1.446	2.262	38
0.6280^h	627.4^h	626.3^h	$0.086\,22^h$	$0.002\,03^h$	0.1971	(0.965)	1.8676	0.4371	274.2^h	1.426	2.124	39
0.6866	686.0	684.9	0.099 30	0.00155	(0.1567)	$0.949^†$	2.3519	0.3471	238.1	1.492	2.171	40
0.615^k	—	—	—	—	0.1893	0.9925	0.8990	0.9081	—	1.659	—	41
0.8850	884.2	883.1	0.088 34	0.001 19	0.2095	$0.929^†$	2.6969	0.3027	267.6	1.014	1.715	42
0.8723	871.6	870.5	0.1057	0.001 06	0.2633	$0.903^†$	3.1812	0.2566	223.7	1.085	1.677	43
0.8721	871.3	870.6	0.1219	0.000 97	0.3031	—	3.6655	0.2227	194.0	1.168	1.721	44
0.8850	884.2	883.1	0.1201	0.000 99	0.3113	—	3.6655	0.2227	196.9	1.218	1.741	45
0.8691	868.3	867.2	0.1223	0.000 97	0.3257	—	3.6655	0.2227	193.4	1.163	1.696	46
0.8661	865.3	864.2	0.1227	0.000 97	0.3214	—	3.6655	0.2227	192.7	1.157	1.708	47
0.9115	910.6	909.5	0.1144	0.001 03	0.1997	—	3.5959	0.2270	206.7	1.133	1.724	48
0.8667	866.0	864.9	0.1390	0.000 97	0.3260	—	4.1498	0.1967	170.4	1.219	1.732	49
0.7967	796.0	794.9	0.040 25	0.001 17	0.5648	—	1.1063	0.7379	587.4	1.352	2.484	50
0.7922	791.5	790.4	0.058 20	0.001 07	0.6608	—	1.5906	0.5132	406.2	1.389	2.348	51
0.7893^m	$786.6^m(34)$	—	$0.035\,52^m$	—	0.0442	0.9995	0.9671	0.8441	—	1.040	—	52
0.8226^h	$821.9^h(35)$	820.8^h	$0.053\,55^h$	—	0.2667	0.9943	1.5195	0.5373	441.6^h	0.8330	—	53
0.7897^h	$789.0^{h,x}(36)$	787.9^h	$0.043\,19^h$	—	0.0920	0.9903	1.1765	0.6939	547.5^h	0.9960	2.08(36)	54
1.397^h	$1396.0^{h,x}(36)$	1395.0^h	$0.045\,89^h$	—	0.2548	$0.9801^†$	2.2117	0.3691	515.3^h	0.6062	1.359(36)	55

(Continued)

TABLE 1.1 (Continued)

Relative density 15°C/15°C[a,b]	kg/m³,[a] (mass in vacuum)	kg/m³,[a,c] (Apparent mass in air)	m³/kmol	Temperature coefficient of density, at 15°C, a −1/°C	Pitzer acentric factor, a	Compressibility factor of real gas, Z 101.3250 kPa (abs), 15°C	Relative density Air=1	Specific volume m³/kg	Volume ratio gas/(liquid in vacuum)	Ideal gas	Liquid	No.
	4. Density of liquid 101.3250 kPa (abs), 15°C			5.	6.	7.	8. Ideal gas* 101.3250 kPa (abs), 15°C			9. Specific heat capacity 101.3250 kPa (abs), 15°C cp kJ/(kg·°C)		
0.6183[h]	617.7[h,x](30)	616.6[h]	0.02757[h]	—	0.2576	0.9899(30)	0.5880	1.388	857.4	2.079	4.693(30)	56
0.856[m](36)	855.0[m]	—	0.0339[m]	—	—	0.9996	1.0000	0.8163	—	1.005	—	57
0.07106[m]	71.00[m](37)	—	0.02839[m]	—	0.219[m]	1.0006	0.0696	11.73	—	14.24	—	58
1.1420[m](25)	1141.0[m](38)	—	0.02804[m]	—	0.0200	0.9993(39)	1.1048	0.7389	—	0.9166	—	59
0.8093[m](26)	806.6[m](31)	—	0.03464[m]	—	0.0372	0.9997	0.9672	0.8441	—	1.040	—	60
1.426	1424.5	1423.5	0.04978	—	0.0737	(0.9875)[+](36)	2.4481	0.3336	475.0	0.4760	—	61
1.000	999.1	998.0	0.01803	0.00014	0.3434	—	0.6220	1.312	1311.0	1.862	4.191	62
0.1251[m]	125.0[m](32)	—	0.03202[m]	—	0	1.0005(40)	0.1382	5.907	—	5.192	—	63
0.8538	853.0[x]	851.9	0.04274	0.006 03	0.1232	—	1.2588	0.6485	553.2	0.7991	—	64

1. Molecular mass (M) is based on the following atomic weights: C = 12.011; H = 1.008; O = 15.9995; N = 14.0067; S = 32.06; Cl = 35.453.

2. Boiling point—the temperature at equilibrium between the liquid and vapor phases at 101.3250 kPa (abs).

3. Freezing point—the temperature at equilibrium between the crystalline phase and the air saturated liquid at 101.3250 kPa (abs).

4. All values for the density and molar volume of liquids refer to the air saturated liquid at 101.3250 kPa (abs), except when the boiling

COMMENTS

Units—all dimensional values are reported in SI units, which are derived from the following basic units:

mass—kilogram, kg

length—meter, m

temperature—International Practical Temperature Scale of 1968 (IPTS-68), where 0°C = 273.15 K

Other derived units are:

volume—cubic meter, m^3

pressure—Pascal, Pa $(1\,Pa = N/m^2)$

Physical constants for molar volume = 22.41383 ± 0.00031
gas constant, $R = 8.31441\,J/(K \cdot mol)$

$$8.31441 \times 10^{-3}\ m^3 \cdot kPa/(K \cdot mol)$$
$$1.98719\ cal/(K \cdot mol)$$
$$1.98596\ Btu(IT)/°R \cdot (lb\text{-}mol)$$

Conversion factors
$1\,m^3 = 35.31467\,ft^3 = 264.1720\,gal$
$1\,kg = 2.204623\,lb$
$1\,kg/m^3 = 0.06242795\,lb/ft^3 = 0.001\,g/cm^3$
$1\,kPa = 0.01\,bar = 9.869233 \times 10^{-3}\,atm$
$= 0.1450377\,lb/in^2$
$1\,atm = 101.3250\,kPa = 14.69595\,lb/in^2$
$= 760\,Torr$
$1\,kJ = 0.2390057\,kcal(thermochemical)$
$= 0.2388459\,kcal(IT)$
$= 0.9478171\,Btu(IT)$

see Rossini, F. D. "Fundamental Measures and Constants for Science and Technology"; CRC Press: Cleveland, Ohio, 1974.

	Field Units	SI Units
P = absolute pressure	psia	kPa
V = volume	ft^3	m^3
n = moles	m/M	m/M
m = mass	lb	kg
M = molecular mass	lb/lb mole	kg/kmole
T = absolute temperature	°R	K
R = universal gas constant	10.73[psia·ft^3/ °R·lb mole mole]	8.3145[kPa m^3/ kmol·K]
ρ = density	slug/ft^3	kg/m^3

A knowledge of the compressibility factor means that the density ρ is also known from the relationship

$$\rho = \frac{PM}{ZRT} \tag{1.2}$$

1.1.1   Compressibility Factor Using the Principle of Corresponding States (CSP)

The following terms are used, P, T, and V, such that

$$P_r = \frac{P}{P_c}, \quad T_r = \frac{T}{T_c}, \quad V_r = \frac{V}{V_c} \tag{1.3}$$

where $P_r, T_r,$ and V_r = reduced parameters of pressure, temperature, and volume

P_c, T_c, and V_c = critical parameters of P, T, and V from Table 1.1

Compressibility factors of many components are available as a function of pressure in most handbooks [1,2]. In application of CSP to a mixture of gases, pseudocritical temperature (T_{pc}) and pressure (P_{pc}) are defined for use in place of the true T_c and P_c to determine the compressibility factor for a mixture.

$$P_{pc} = \sum y_i P_{ci}, \quad T_{pc} = \sum y_i T_{ci} \tag{1.4}$$

$$P_{pr} = P/P_{pc}, \quad T_{pr} = T/T_{pc} \tag{1.5}$$

where subscript i = component in the gas mixture
y_i = mole fraction of component "i" in the gas mixture

For given values of P_{pr} and T_{pr}, compressibility factor Z can be determined from Figure 1.1 [1]. If a gas mixture contains significant concentrations of carbon dioxide and hydrogen sulfide, then corrected pseudocritical

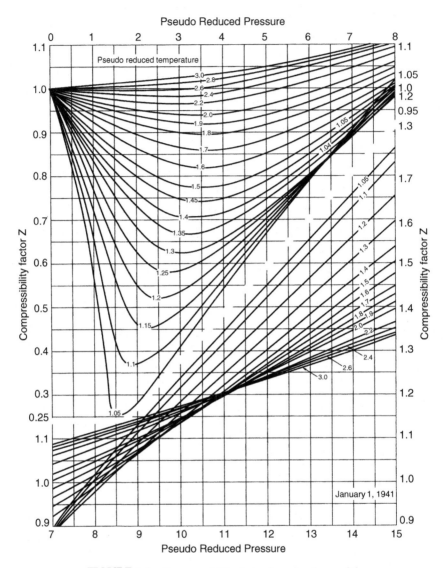

FIGURE 1.1   Compressibility factor for natural gases [1].

constants (T'_{pc}, P'_{pc}) are defined as follows:

$$T'_p c = T_{pc} - e \tag{1.6}$$

$$P'_{pc} = \frac{P_{pc} T'_{pc}}{T_{pc} + y_{H_2S}(1 - y_{H_2S})e} \tag{1.7}$$

where

$$e = 120 \left[\left(y_{CO_2} + y_{H_2S} \right)^{0.9} - \left(y_{CO_2} + y_{H_2S} \right)^{1.6} \right] + 15 \left(y_{H_2S}^{0.5} - y_{H_2S}^{4.0} \right) \tag{1.8}$$

where y_{CO_2}, y_{H_2S} = mole fraction of CO_2 and H_2S in mixture.

1.1.2 Direct Calculation of Z Factors

The Hall-Yarborough equation is one of the best:

$$Z = \frac{0.06125 P_{pr} t \exp[-1.2(1-t)^2]}{y} \tag{1.9}$$

where P_{pr} = the pseudoreduced pressure

t = the reciprocal, pseudoreduced temperature T_{pc}/T

y = the reduced density which can be obtained as the solution of the equation

$$-0.06125_{pr} t \exp \left[-1.2(1-t)^2 \right] + \frac{y + y^2 + y^3 - y^4}{(1-y)^3}$$

$$- \left(14.76t - 9.76t^2 + 4.58t^3 \right) y^2$$

$$+ \left(90.7t - 24.2t^2 + 42.4t^3 \right) y^{(2.18+2.82t)} = 0 \tag{1.10}$$

This nonlinear equation can be conveniently solved for y using the simple Newton-Raphson iterative technique.

Good results give cubic equation of state and its modifications. Most significant of the modifications are those of Soave and Peng and Robinson [5].

The Soave equation (SRK) is [4]:

$$P = \frac{RT}{V-b} - \frac{a}{V(V+b)} \tag{1.11}$$

$$Z^3 = Z^2 + \left(A - B - B^2 \right) Z - AB = 0 \tag{1.12}$$

where $b = \Sigma x_i b_i$ for a mixture

$b_i = 0.08664 RT_{ci}/P_{ci}$ for a single component

$a = \Sigma_i \Sigma_j x_i x_j (a_i a_j)^{0.5} (1 - k_{ij})$ for a mixture

$a_i = a_{ci} \alpha_i$ for a single component

$a_{ci} = 0.42748 (RT_{ci})^2 / P_{ci}$

$$\alpha_i^{0.5} = 1 + m_i\left(1 - T_{ri}^{0.5}\right)$$
$$m_i = 0.48 + 1.574\omega_i - 0.176\omega_i^2$$
$$A = aP/(RT)^2, \quad B = bP/RT$$

The Peng-Robinson equation (PR) is

$$P = \frac{RT}{V - b} - \frac{a}{V(V+b) + b(V-b)} \tag{1.13}$$

$$Z^3 - (1-B)Z^2 + \left(A - 2B - 3B^2\right)Z - \left(AB - B^2 - B^3\right) = 0 \tag{1.14}$$

where $b = \Sigma x_i b_i$
$\quad b_i = 0.077796 RT_{ci}/P_c$
$\quad a = \Sigma_i \Sigma_j x_i x_j (a_i a_j)^{0.5}(1 - k_{ij})$
$\quad a_i = a_{ci}\alpha_i$
$\quad a_{ci} = 0.457237(RT_{ci})^2/P_{ci}$
$\quad \alpha_i^{0.5} = 1 + m_i\left(1 - T_{ri}^{0.5}\right)$
$\quad m_i = 0.37646 + 1.54226\omega_i - 0.26992\omega_i^2$
$\quad$ A, B as in SRK equation

where P = pressure (absolute units)
$\quad T$ = temperature (°R or K)
$\quad R$ = universal gas constant
$\quad Z$ = compressibility factor
$\quad \omega$ = acentric Pitzer factor (see Table 1.1)
T_{ci}, P_{ci} = critical parameters (see Table 1.1)
$\quad k_{ij}$ = interaction coefficient ($= 0$ for gas phase mixture)

Both SRK and PR equations are used to predict equilibrium constant K value. See derivation of vapor-liquid equilibrium by equation of state at end of this chapter.

Reservoir hydrocarbon fluids are a mixture of hydrocarbons with compositions related to source, history, and present reservoir conditions. Consider the pressure-specific volume relationship for a single-component fluid at constant temperature, below its critical temperature initially hold in the liquid phase at an elevated pressure. This situation is illustrated in Figure 1.2 [1]. Bubble point and dew point curves in Figure 1.2a correspond to the vapor pressure line in Figure 1.2b. A locus of bubble points and a locus of dew points that meet at a point C (the critical point) indicate that the properties of liquid and vapor become indistinguishable [6].

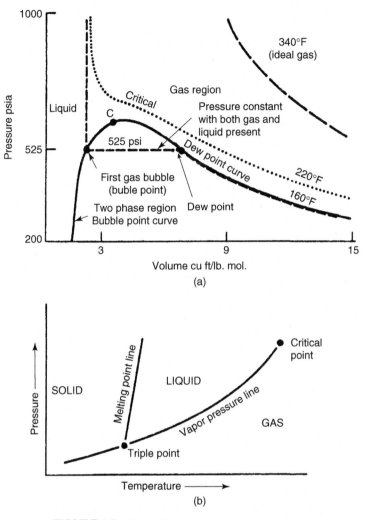

FIGURE 1.2   Phase diagram for a single component.

Multicomponent systems have different phase behavior than pure component. In the P–T diagram instead of the vapor-pressure line, we have an area limited by saturation line (bubble point+dew point), see Figure 1.3 [1]. The diagram's shape is more or less the same for two- or three-component systems as for multicomponent systems.

For isothermal production in the reservoir, position A indicates reservoir fluid found as an underposition saturated oil, B indicates reservoir fluid found as a gas condensate, and position C indicates reservoir fluid found as a dry gas. Expansion in the liquid phase to the bubble point at

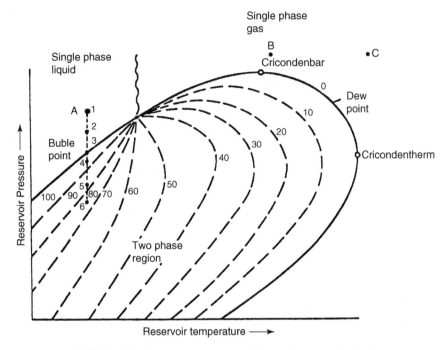

FIGURE 1.3   Pressure–temperature diagram for reservoir fluids.

constant temperature is similar to a pure component system. Expansion through the two-phase region does not occur at constant pressure, but is accompanied by a decrease in pressure as the composition of liquid and vapor changes.

1.1.3 Classification of Hydrocarbon Fluids

Hydrocarbon fluids usually are classified as to the phase behavior exhibited by the mixture. Figure 1.4 shows the pressure–temperature phase diagram of the four general classifications of fluids: dry gas, gas condensate, volatile oil, and black oil. As it can be seen, the source temperature also plays a role in the determination of fluid type. According to MacDonald [7], each type of fluid has composition as given in Table 1.2. Sometimes hydrocarbon mixtures are classified as follows: dry gas, wet gas, gas condensate, and black oil (Figure 1.4).

In the case of dry gas, a light hydrocarbon mixture existing entirely in gas phase at reservoir conditions and a decline in reservoir pressure will not result in the formation of any reservoir liquid phase; it is a rather theoretical case. Usually gas reservoirs fall into the next group, wet gas.

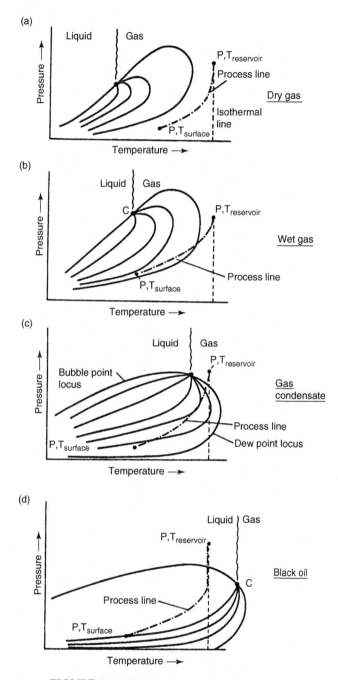

FIGURE 1.4   Hydrocarbon mixture classification.

TABLE 1.2 Typical Compositions of Hydrocarbons Fluids [7]

	Dry Gas	Wet Gas	Retrograde Condensate	Volatile Oil or Near Critical Oils	Black Oil	Heavy Oil
C_1 (mole %)	90		70	5	30	
C_2–C_6 (mole %)	9		22	30	35	
C_{7+} (mole %)	1		8	15	35	
GOR [ft^3/bbl]	∞	150,000	3000–150,000 (10,000+)*	2000–3000 (3000–6000)*	<2000	
[m^3/m^3]	∞	>27,000	540–27,000	360–540	<360	
API Liquid Gravity	—		40–60	45–70	30–45	<20
Liquid Specific Gravity	—		0.83–0.74	0.8–0.7	0.88–0.8	>0.94

Gas condensate or retrograde gas system is the case when the critical temperature of system is such that reservoir temperature is between critical and cricondentherm as shown in Figure 1.4c. If the pressure is reduced to the cricodenbar pressure, the liquid phase is increasing, but the liquid phase may reevaporate later on. This phenomena — the condensation of liquid upon decrease in pressure — is termed isothermal retrograde condensation. The liquid phase recovered from a condensate system is recovered from a phase that is vapor at reservoir conditions. This is also partly true of volatile oil systems where the vapor phase in equilibrium with the reservoir liquid phase is particularly rich in liquefiable constituents (C_3 to C_{8+}), and a substantial proportion of stock tank liquid may derive from a reservoir vapor phase. We normally do not expect to see retrograde behavior at reservoir pressures below about 2500 psi (17.2 MPa).

Volatile oil systems are those within the two-phase region under reservoir conditions, the vapor stage corresponding to condensate compositions and conditions. Volatile oil is not an apt description because virtually all reservoir fluids are volatile. What is really meant is that the reservoir fluid exhibits the properties of an oil existing in the reservoir at a high temperature near its critical temperature. These properties include a high shrinkage immediately below the bubble point. In extreme cases, this shrinkage can be as much as 45% of the hydrocarbon pore space within 10 psi (0.7 bar) below the bubble point. Near-critical oils have formation factor B_o of 2 or higher; the compositions are usually characterized by 12.5 to 20 mole % heptanes or more.

Ordinary oils are characterized by GOR from 0 to approximately 200 ft^3/bbl (360 m^3/m^3) and with B_0 less than 2. Oils with high viscosity (about 10 cp), high oil density, and negligible gas/oil ratio are called heavy oils. At surface conditions may form tar sands. Oils with smaller than 10 cp viscosity are known as a black oil or a dissolved gas oil system; no anomalies are in phase behavior. There is no sharp dividing line between each group of reservoir hydrocarbon fluid; however, liquid volume percent versus pressure diagram Figure 1.5a–e is very useful to understand the subject.

The significant point to be made is that when an oil system exists in intimate contact with an associated gas cap, the bubble point pressure of the oil will be equal to the dew point pressure of the gas cap, and both those values will be equal to the static reservoir pressure at the gas–oil contact, Figure 1.6.

1.1.4 Reservoir Conditions Phase Behavior

There is one phase flow in reservoir conditions if well flowing pressure P_{wf} is higher than bubble point pressure P_b or dew point pressure P_d and two-phase flow occurs by a wellbore. Reservoir depletion and production

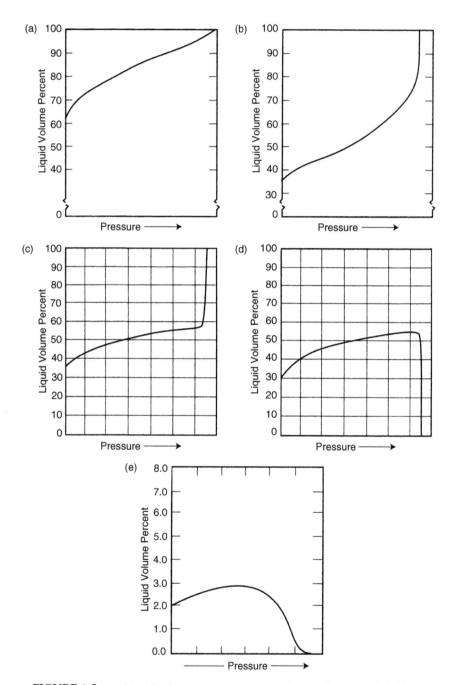

FIGURE 1.5a–e   Liquid volume percent — pressure diagram for reservoir fluid [6].

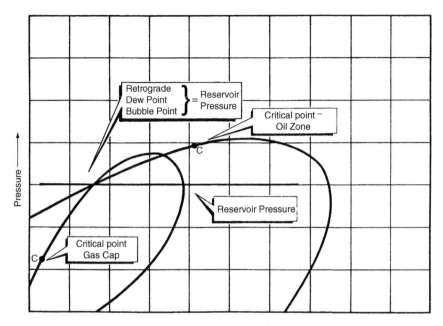

FIGURE 1.6   Pressure–temperature diagram for gas cap and associated oil [6].

consist of two separate processes: flash liberation (vaporization) and differential liberation. A schematic representation of flash vaporization is shown in Figure 1.7. At stage 1 reservoir fluid is under reservoir pressure and temperature at known volumes V_t. The pressure in the cell is covered by increasing the space available in the cell for the fluid V_{t2}. This procedure is repeated until a large change in the pressure–volume slope is indicated. The above procedure indicates that flash vaporization is a phase-changing process due to change in pressure and temperature if the mass of reservoir fluid or total composition system remains constant and can be expressed as follows:

$$zF = xL + yV \tag{1.15}$$

where

z = mole fraction of component in a reservoir fluid mixture
F = number of moles of sample at initial reservoir pressure and temperature
x = mole fraction of component in liquid (e.g., P_5)
L = number of moles of equilibrium liquid
y = mole fraction component in gas mixture
V = number of moles of equilibrium gas phase

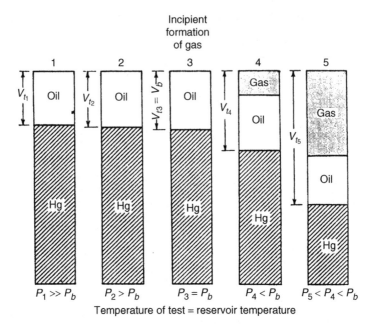

FIGURE 1.7 Schematic representation of laboratory P–V equilibrium (flash vaporization) [10].

Equilibrium or flash liberation calculations may be made for reservoir fluid that divides into two phases at any temperature and pressure.

Schematic representation of differential liberation in laboratory conditions is shown in Figure 1.8. It begins in the same manner as flash vaporization. The sample is placed in a pressure higher than bubble point pressure. The pressure is lowered until such time that free gas is liberated in the cell. Then for predetermined pressure or volume increments, mercury is withdrawn from the cell, gas is released from solution, and the cell is agitated until the liberated gas is in the equilibrium with the oil. All free gas is ejected from the cell at a constant pressure by injection of mercury. The volume of free gas is displaced, and the oil remaining in the cells are thus measured at cell conditions. This procedure is repeated for all the pressure increments until only oil remains in the cell at reservoir temperature and atmospheric pressure.

In contrast to flash vaporization, differential vaporization is undertaken with the decreasing mass participating in the process. Equation 1.15 cannot be applied because F is not constant anymore at the given set pressure and temperature.

The question arises: Which process occurs in reservoir conditions? Some specialists, e.g., Moses [8], assume that the reservoir process is a combination of differential and flash. Such statements are incorrect. They produce

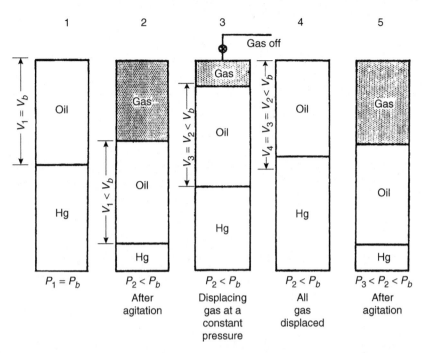

FIGURE 1.8 Schematic representation of laboratory differential vaporization [10].

misunderstanding and confusion. The following is meant to clarify and straighten out this problem:

1. In reservoir conditions, differential process always occurs.
2. In production tubing and surface pipeline flow and in the separators, the flash process takes place (subject to some limiting assumption).

Flash process refers to the conditions in which the mass of the considered system does not vary with changes in pressure and temperature. Flash process in two-phase regions (vaporization or condensation) can be defined in terms of total system composition. The total system composition (z_i) can be measured at any point outside of saturation line, e.g., points A, B, and C (Figure 1.3). As a substitution the following treatment can be used; the total system composition in two-phase region flash process remains constant. The flash process may ensue for a composition z_i that separates into two phases for the values of pressures and temperatures inside the saturation curve area. After the temperature and pressure are chosen, all the gas is in equilibrium with all the oil. In other words, a change of pressure or temperature, or both, in a flash process can change the equilibrium conditions according to the Gibbs phase rule. This rule provides the number of independent variables that, in turn, define intensive properties. Flash vaporization may be a batch or continuous process. Treating two-phase flow in tubing as a steady state, neglecting the gas storage effect, and gas slippage result in flash process. In a horizontal flow, and in separators, a similar flash process comes about.

The same kind of equilibrium, but with its fluid mass decreasing differentially, is called a *differential process* (liberation or condensation).

In reservoir conditions the hydrocarbon pore volume (HCPV) remains constant if the expansion of interstitial water and rock compressibility are neglected. For such constant HCPV, it must be made clear that differential process occurs always as differential vaporization of differential condensation. Differential vaporization takes place when the reservoir temperature is less than critical temperature of solution ($T_{res} < T_c$), and also it takes place during retrograde gas reservoir depletion, but only in the region pressure and temperature at which the retrograde liquid is vaporized.

In differential condensation, the oil reservoir pressure is maintained constant or almost constant — for example, by gas injection. Differential condensation can also occur just below the dew point in a gas-condensate reservoir.

Above the bubble point and the dew point curves, the virtual (apparent) value of vaporization and/or condensation is zero, but because the mass of fluid in a depleted reservoir is changing as a result of decreasing pressure, the process could be assumed to be differential. One important statement has to be added: there is no qualitative difference between the reservoir fluid in either differential or flash process, if pressure and temperature fall into the area outside of the saturation curves (Figure 1.3). A schematic representation of differential vaporization of oil in reservoir conditions is shown in Figure 1.9. As indicated in Figure 1.9, six hypothetical cases are distinguished. Study Figure 1.9 simultaneously with Figure 1.3 and Figure 1.10.

Consider a first sample, in which there is a fixed mass of oil at given temperature, pressure, and HCPV. When the pressure P_1 drops to P_2, the volume of oil increases but the HCPV does not change, so the difference in oil-removed volume equals the total oil production when the pressure changes from P_A to P_2 (sample 2).

The third sample is considered at the bubble point; the oil volume change between P_2 and P_b resembles that between P_1 and P_2. Beginning at P_b, the first gas bubbles are released. Pressure P_4 corresponds to the lowest value of GOR (Figure 1.10) and coincides with the highest pressure in a two-phase region, in which only one phase (oil) still flows. Pressure P_4 could be called a *gas flow saturation pressure*. Between P_b and P_4, compositions x_i, y_i, and z_i are changed. HC mass in pore volume is decreasing, so it is the differential process that is contrary to Moses' [8] belief that this is a flash process.

At point 5, HCPV remains constant as in steps 1 to 4, the oil volume has changed and the system is into a two-phase region. An amount of released gas exceeds the gas flow saturation pressure P_4; gas begins to

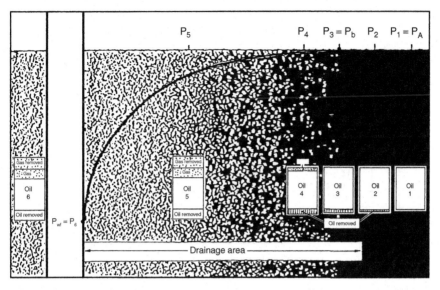

FIGURE 1.9   Schematic representation of differential vaporization in reservoir conditions.

run and is partially removed from the HCPV. This is how two-phase flow is generated. Sample 6 characterizes the same process very close to the bottom-hole area. The reservoir fluid mass difference in steps 1 and 6 equals the total production from an HCPV.

In conclusion, it has been shown that the flash process occurs whenever we are dealing with a closed system or a steady-state flow, e.g., a two-phase flow in vertical tubing, in horizontal pipe flow, and in separators. For any open system, such as a reservoir formation, or for an unsteady-state flow, the differential process is properly describing the quasiequilibrium conditions.

1.2 SAMPLING PROCESS AND LABORATORY MEASUREMENTS

The overall quality of the reservoir fluid study and the subsequent engineering calculations based upon that study can be no better than the quality of the fluid samples originally collected during the field sampling process.

Samples representative of the original reservoir can be obtained only when the reservoir pressure is equal to or higher than the original bubble point or dew point.

The pressure drawdown associated with normal production rates will cause two-phase flow near the wellbore if the fluid in the formation was initially saturated or only slightly undersaturated. Relative permeability

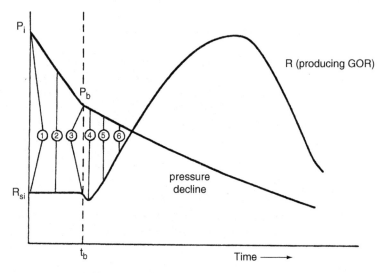

	Sample #					
Reservoir parameters	1	2	3	4	5	6
HCPV	$v_1 = v$	$v_2 = v$	$v_3 = v$	$v_4 = v$	$v_5 = v$	$v_6 = v$
HC Mass in Pore Volume (m)	$m_1 = m$	$m_2 < m_1$	$m_3 < m_2$	$m_4 < m_3$	$m_5 < m_4$	$m_6 < m_5$
Pressure (P)	$P_1 = P_A$	$P_2 < P_1$	$P_b < P_2$	$P_4 < P_b$	$P_5 < P_4$	$P_6 < P_5$
Temperature (T)			assumed constant reservoir temperature			
Total Reservoir Composition z_i	$z_{i1} - z_{iA}$	$z_{i2} - z_{i1}$	$z_{ib} - z_{i1}$	$z_{i4} - z_{i1}$	$z_{i5} \neq z_{i1} \neq z_{i4}$	$z_{i6} \neq z_{i1} \neq z_{i4} \neq z_{i5}$
Gas Composition y_i	$y_{i1} = y_{iA} = 0$	$y_{i2} = y_{iA} = 0$	$y_{i3} = y_{ib} \neq 0$	$y_{i4} = y_{i1} \neq y_{i3}$	$y_{i5} = y_{i1} \neq y_{i3} \neq y_{i4}$	$y_{i6} = y_{i1} \neq y_{i3} \neq y_{i4} \neq y_{i5}$
Oil Composition x_i	$x_{i1} = z_{iA} = z_{i1}$	$x_{i2} = x_{i1} = z_{i1}$	$x_{i3} = x_{ib} \neq x_{i1}$	$x_{i4} \neq x_{i1} \neq x_{i3}$	$x_{i5} \neq x_{i1} \neq x_{i3} \neq x_{i4}$	$x_{i6} \neq x_{i1} \neq x_{i3} \neq x_{i4} \neq x_{i5}$

FIGURE 1.10   Schematic representation reservoir pressure (P) and GOR vs. time and mass and composition in reservoir differential vaporization process.

effects may then cause the material entering the wellbore to be different from the original overall (total) composition fluid existing at the boundary of drainage area. The problem of drawdown in a saturated reservoir cannot be eliminated; therefore, it is necessary to limit the pressure drawdown by reducing the flowrate to the lowest possible stable rate while sampling.

There are two basic methods of sample collection: subsurface (bottom-hole) and surface (separator). The fluid sampling method to be used dictates the remainder of the conditioning process. If the bottomhole samples are to be collected, the period of reduced flowrate will generally last from 1 to 4 days, depending on the formation and fluid characteristics and the drainage area affected. After this reduced flowrate period, the well would be shut in and allowed to reach static pressure. The shut-in period would last about 1 day or up to a week or more, depending on formation characteristics. For the case of the saturated reservoir, the shut-in period has the resultant effect of forcing gas into solution in the oil, thus raising the saturation pressure. In some cases, the desired value of P_b is obtained; however, in most cases this value is only approached and the final difference is a function of well productivity, production rate, and fluid properties. At the conclusion of the shut-in period, the well would be properly conditioned and ready for bottomhole sampling. Subsurface sampling is generally not recommended for gas-condensate reservoirs; the same is true for oil reservoirs producing substantial quantities of water. If separator gas and liquid samples are to be collected, the gas and liquid rates must be monitored continually during the period of stable flow at reduced flowrates. A minimum test of 24 hours is recommended, but more time may be needed if the pressure drawdown at the formation has been high. Surface sampling, called *separator sampling*, has wider applications than subsurface sampling, and is the only recommended way of sampling a gas-condensate reservoir, but often can be used with good success for oil reservoirs as well. There are three requirements to successful separator sampling.

1. stable production at a low flowrate
2. accurate measurement of gas and oil flowrates
3. collection of representative samples of first-stage gas and first-stage liquid

The above procedure is described in detail in API Standard 811-08800 [9].

The reservoir process is stimulated in the laboratory by flash differential vaporization (Figure 1.7 and Figure 1.8). Based on both figures, it is possible to prepare the reservoir fluid data for engineering calculations.

In the laboratory, the differential liberation consists of a series — usually 10 to 15 — of flash liberations. An infinite series of flash liberations is the equivalent of a true differential liberation. At each pressure level, gas is evolved and measured. The volume of oil remaining is also measured at each depletion pressure. This process is continued to atmospheric pressure. The oil remaining at atmospheric pressure is measured and converted to a volume at 60°F (15.6°C). This final volume is referred to as the residual oil. The volume of oil at each of the higher pressures is divided by the volume of residual oil at 60°F (15.6°C).

Example 1.1

Surface separator samples were collected from a well on completion of a 2-hr test on June 8, 1984. The gas/liquid ratio measured on this test was 4565 ft.^{3} of separator gas per barrel of separator liquid and was used as the basis for this recombination. The resultant reservoir fluid exhibited a dew point of 4420 psia at $T_{res} = 285°F$. The reservoir fluid exists as a gas (an undersaturated gas) at $P_{res} = 12,920$ psia.

A constant volume depletion study is also performed on the reservoir fluid. The produced compositions and volumes from the depletion study are used in conjunction with equilibrium constants to calculate cumulative STO and separator gas recoveries resulting from conventional field separation. Gas plant products in both the primary separator gas and the full well stresses should also be reported.

Sampling Conditions

Date sampled (on 20/64 choke)	06-08-84 for 1330 hr
Tubing pressure, flowing	9960 psig
Primary separator temperature	95°F
Primary separator pressure	900 psig
Primary separator gas rate (Table 1.3)	2229.7 MCF/Day
Liquid rate (2nd stage @ 50 psig)	396 bbl/day
Gas/liquid ratio (GOR)	5631 SCF 1st stg. gas/bbl 50 lb liq.
Shrinkage factor (vol. 50 lb liq./vol. sep. liq.)	0.8108
Gas/liquid ratio (GOR)	4565 SCF 1st stg. gas/bbl 900 lb liq.
Shrinkage factor (vol. S.T. liq./vol. sep. liq.) through 50-lb 2nd stage	0.7445
Pressure base	15.025 psia @ 60°F

The samples of separator gas and separator liquid were analyzed and the results are reported in Table 1.4, showing both composition of each sample and the computed analysis of the well stream based on the GOR in the primary separator. The separator liquid (oil) production was calculated from the measured second-stage production by applying the determined shrinkage factor.

TABLE 1.3   Calculation of Gas Rate [11]

$\sqrt{HwPf} = 165.6804$	$Hw = 30.0'' \, H_2O,$	$Pf = 915.00 \, psia$
$Fb = 455.0300$	$D = 5.761'', 15.025 \, psia$	$d = 1.50''$
$Fpb = 0.9804$		
$Fr = 1.0002$	$b = 0.0367$	
$Y2 = 1.0002$	$Hw/Pf = 0.033,$	$d/D = 0.260$
$Fg = 1.1953$	$Gravity = 0.6999,$	$Fg = \sqrt{1/0.6999}$
$Ftf = 0.9680$	$Temp. = 95°F,$	$Ftf = \sqrt{520/555}$
$Fpv = 1.0859$	$pTr' = 1.441,$	$pPr' = 1.372$
	$Z = 0.8480,$	$Fpv = \sqrt{1/Z}$

Acid Gas Correction Factor   Epsilon = 3.57

$Q = \sqrt{Hw\,Pf} \times Fb \times Fpb \times Fr \times Y2 \times Fg \times Ftf \times Fpv \times 24$

$Q = 2229.7$ MCF/day@15.025   PSIA@60°F

TABLE 1.4   Hydrocarbon Analysis of Separator Products and Calculated Wellstream [11]

Component	Separator Liquid Mol %	Separator Liquid Liq. Vol. %	Separator Gas Mol %	Separator Gas GPM@ 15.025 PSIA	Well Stream Mol %	Well Stream GPM@ 15.025 PSIA
Carbon dioxide	0.41	0.19	2.18		1.84	
Nitrogen	0.00	0.00	0.16		0.13	
Methane	22.36	10.16	82.91		71.30	
Ethane	8.54	6.12	8.22	2.243	8.28	2.259
Propane	10.28	7.58	3.82	1.073	5.06	1.421
Iso-Butane	5.69	4.99	1.11	0.372	1.99	0.664
N-Butane	5.11	4.32	0.80	0.256	1.63	0.523
Iso-Pentane	4.84	4.75	0.30	0.113	1.17	0.437
N-Pentane	2.01	1.95	0.17	0.064	0.52	0.193
Hexanes	8.16	8.75	0.22	0.091	1.74	0.731
Heptanes Plus	32.60	51.19	0.11	0.053	6.34	3.907
Total	100.00	100.00	100.00	4.265	100.00	10.135

Calculated specific gravity (Air = 100)= 0.6999 separator gas 1.0853 well stream
Sep. gas heat of combustion (BTU/Cu.ft. @ 15.025 PSIA & = 1206.8 real 60°F) dry
Sep. gas heat of combustion (BTU/Cu.ft. @ 15.025 PSIA & = 1185.7 water sat. 60°F) wet
Sep. gas compressibility (@ 1 ATM. & 60°F) Z = 0.9968

Properties of Heptanes Plus: Properties of Separator Liquid:

Specific gravity = 0.7976   0.6562 @ 60/60°F
Molecular Weight = 152   78.7
Cu.Ft./Gal. = 16.24   25.81 @ 15.025 PSIA & 60°F

Properties of Stock Tank Liquid:

Gravity = 56.8 degrees API @ 60°F

Basis of Recombination:

Separator liquid per MMSCF separator gas = 219.05 bbls

1.2.1   Equilibrium Cell Determinations

Following the compositional analyses, portions of the primary separator liquid and gas were physically recombined in their produced ratio in a variable volume, glass-windowed equilibrium cell. Determinations on this mixture were divided into the following main categories.

1. Dew point pressure determination and pressure–volume relations on a constant weight of reservoir fluid at the reservoir temperature: the procedure consisted of establishing equilibrium between gas and liquid phases at a low pressure and measuring volumes of liquid and gas in

equilibrium at that pressure. The pressure was raised by the injection of mercury into the cell and phase equilibrium established again at a higher pressure. This procedure was repeated until all of the liquid phase had vaporized, at which point the saturation pressure was observed. The cell pressure was then raised above the dew point pressure in order to determine the supercompressibility characteristics of the single-phase vapor. As a check on all readings and, particularly, to verify the dew point, the cell pressure was incrementally reduced, equilibrium established and volumetric readings made. Reported in Table 1.5 are the relative volume relations (Figure 1.11) and specific volumes of the reservoir fluid over a wide range of pressures as well as compressibility factors (Figure 1.12), the single-phase vapor above the dew point. Reported in Table 1.6 are the dew point pressure (Figure 1.13) resulting from recombination's at gas/liquid ratios above and below the ratio measured at the time of sampling.

2. Compositions of the produced wellstream and the amount of retrograde condensation resulting from a stepwise differential depletion. This procedure consisted of a series of constant composition expansions and constant pressure displacements with each displacement being terminated at the original cell volume. The gas removed during the constant pressure was analyzed. The determined compositions (Figure 1.14), computed GPM content (Figure 1.15), respective compressibility (deviation) factors, and volume of wellstream produced during depletion (Figure 1.16) are presented in Table 1.7. The volume of

TABLE 1.5   Pressure–Volume Relation of Reservoir Fluid at 285°F [11]

Pressure (PSIA)	Relative Volume (V/Vsat)	Specific Volume (Cu.Ft./Lb.)	Retrograde BPMMCF*	Liquid Vol. %**	Deviation Factor (Z)	Calculated Viscosity (Centipoise)
***12,920 Res.	0.6713	0.03635			1.8485	0.0793
10000	0.7201	0.03898			1.5346	0.0674
8000	0.7741	0.04191			1.3197	0.0579
6000	0.8639	0.04677			1.1047	0.0471
5000	0.9373	0.05075			0.9988	0.0412
4420 D.P.	1.0000	0.05414	0.00	0.00	0.9420	0.0374
					(Two Phase)	
4000	1.0677	0.05781	10.58	1.30	0.9102	
3500	1.1764	0.06369	69.83	8.55	0.8775	
3000	1.3412	0.07261	94.17	11.53	0.8575	
2500	1.5879	0.08597	107.92	13.21	0.8460	
2000	1.9831	0.10737	114.27	13.99	0.8453	
1500	2.6605	0.14404	114.27	13.99	0.8505	
1000	4.0454	0.21902	107.92	13.21	0.8621	

*BBLS. per MMSCF of dew point fluid.
**Percentage of hydrocarbon pore space at dew point.
***Extrapolated.

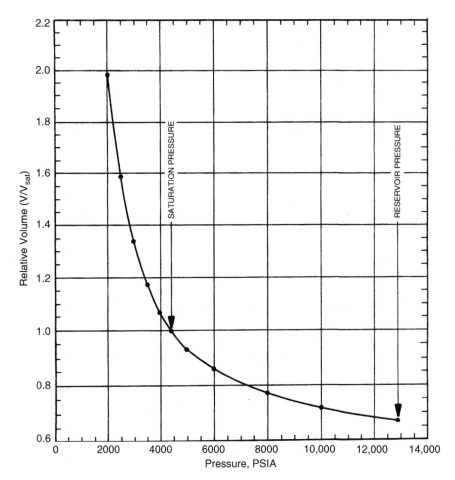

FIGURE 1.11   Pressure–volume relation of reservoir fluid at 285°F (constant mass expansion) [11].

retrograde liquid resulting from the gas depletion is shown in Table 1.8 (and Figure 1.17), both in terms of barrels of reservoir liquid and percentage of hydrocarbon pore space. Show in Table 1.7 are the compositions of the gas and liquid remaining in the reservoir after depletion to abandonment pressure.

1.2.2 Equilibrium Flash Calculations

The produced compositions and volumes from a depletion study were used in conjunction with equilibrium constants K (derived from a Wilson modified R-K equation of state) to calculate cumulative stock tank liquid and separator gas recoveries resulting from conventional separation. These

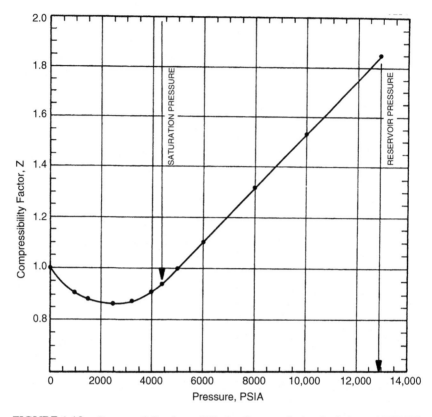

FIGURE 1.12 Compressibility factor "Z" of wellstream during depletion at 285°F [11].

TABLE 1.6 Observed Saturation Pressures From Stepwise Recombination at 285°F [11]

Gas-Liquid Ratio (SCF 1st Stg. Gas) (BBL. 1st Stg. Liq.)	Saturation Pressure (Psia)
9000	6000 Dew Point
4565 (Produced)	4420 Dew Point
2500	3830 Dew Point

data are reported in Tables 1.9 and 1.10. Also, gas plant products in both the primary separator and the full wellstream are attached.

Example 1.2
This is a black oil problem. From differential vaporization (Table 1.11) and separator test data (Table 1.12) discuss the B_o and R_s calculation method.

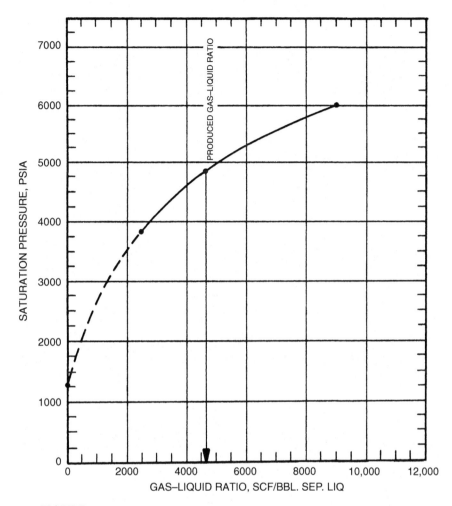

FIGURE 1.13   Effect of gas/liquid ratio upon saturation pressure at 285°F [11].

Figure 1.19 and Figure 1.20 illustrate laboratory data. These data are reported in a convention other than in Example 1.1. The residual oil in the reservoir is never at 60°F (15.6°C), but always at T_{res}. Reporting these data relative to the residual oil at 60°F (15.6°C), gives the relative-oil-volume curve the appearance of an FVF curve, leading to its misuse in reservoir calculations. A better method of reporting these data is in the form of a shrinkage curve. We may convert the relative-oil-volume data in Figure 1.19 and Table 1.11 to a shrinkage curve by dividing each relative-oil-volume factor B_{od} by the relative oil volume factor at the bubble point, B_{odb}.

The shrinkage curve now has a value of one at the bubble point and a value of less than one at the subsequent pressures below the bubble point,

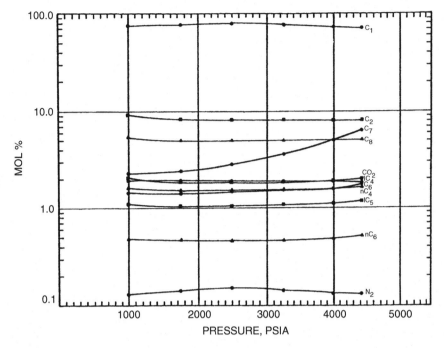

FIGURE 1.14 Hydrocarbon composite of wellstreams produced during pressure depletion [11].

as in Figure 1.21. As pressure is reduced and gas is liberated, the oil shrinks. The shrinkage curve describes the volume of this original barrel of oil in the reservoir as pressure declines. It does not relate to a stock tank or surface barrel.

We now know the behavior of the oil in the reservoir as the pressure declines. We must have a way of bringing this oil to the surface through separators and into a stock tank. This process is a flash process. Most reservoir fluid studies include one or more separator tests to simulate this flash process. Table 1.12 is a typical example of a set of separator tests. During this test, the FVF is measured. The FVF is the volume of oil and dissolved gas entering the wellbore at reservoir pressure and temperature divided by the resulting stock-tank oil volume after it passes through a separator.

The FVF is B_o; because separators result in a flash separation, we showed a subscript, B_{of}. In most fluid studies, these separator tests are measured only on the original oil at the bubble point. The FVF at the bubble point is B_{ofb}. To make solution-gas-drive or other material-balance calculations, we need values of B_{of} at lower reservoir pressures. From a technical standpoint, the ideal method for obtaining these data is to place a large sample of reservoir oil in a cell, heat it to reservoir temperature, and pressure-deplete it with a differential process to stimulate reservoir depletion. At

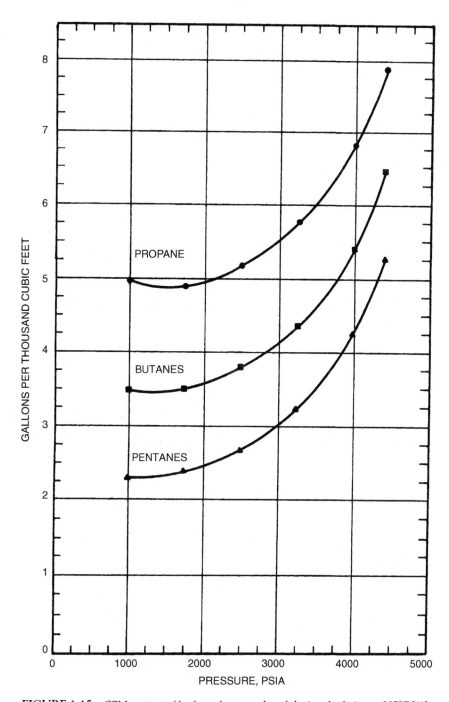

FIGURE 1.15 GPM content of hydrocarbons produced during depletion at 285°F [11].

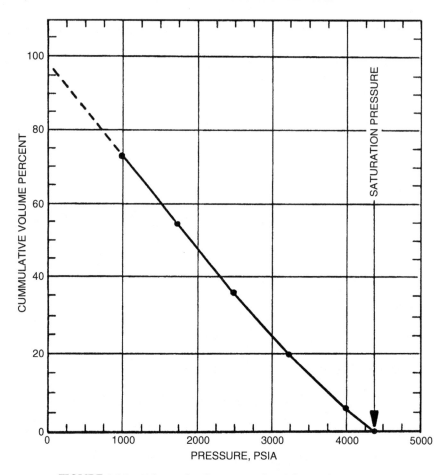

FIGURE 1.16   Volume of wellstream produced during depletion [11].

some pressure a few hundred psi below the bubble point, a portion of the oil is removed from the cell and pumped through a separator to obtain the flash FVF, B_{of}, at the lower reservoir pressure. This should be repeated at several progressively lower reservoir pressures until a complete curve of B_{of} versus reservoir pressure has been obtained. The process is time consuming and consequently adds to the cost of the study. Most studies include only values of B_{ofb}, the FVF at the bubble point. The values of B_{of} at lower pressures must be obtained by other means.

The method calls for multiplying the flash FVF at the bubble point B_{ofb} by the shrinkage factors at various reservoir pressures obtained earlier. The shrinkage factor was calculated by dividing the relative oil volume factors B_{od} by the relative oil volume factor at the bubble point B_{ofb}. If we combine both calculations, we can start with the differential-relative-volume curve

TABLE 1.7 Depletion Study at 285°F [11]

Component	Dew Point Fluid 4420	Reservoir Pressure — PSIA					Aband. Liquid 1000
		4000	3250	2500	1750	1000	
Nitrogen	0.13	0.13	0.14	0.15	0.14	0.13	0.04
Carbon dioxide	1.84	1.87	1.90	1.92	1.94	1.93	0.67
Methane	71.30	73.09	74.76	75.69	76.09	74.71	16.61
Ethane	8.28	8.18	8.14	8.12	8.15	9.05	4.76
Propane	5.06	5.01	4.97	4.94	4.93	5.29	4.60
Iso-butane	1.99	1.93	1.89	1.87	1.89	2.02	2.66
N-Butane	1.63	1.59	1.56	1.55	1.54	1.61	2.37
Iso-pentane	1.17	1.12	1.09	1.06	1.06	1.10	2.37
N-Pentane	0.52	0.48	0.47	0.46	0.47	0.48	1.17
Hexanes	1.74	1.59	1.51	1.45	1.40	1.43	5.96
Heptanes Plus	6.34	5.01	3.57	2.79	2.39	2.25	58.77
Totals	100.00	100.00	100.00	100.00	100.00	100.00	100.00
Well stream gravity (Air = 1)	1.0853	1.0043	0.9251	0.8823	0.8595	0.8606	

(Continued)

TABLE 1.7 (Continued)

Component	Dew Point Fluid 4420	Reservoir Pressure—PSIA						Aband. Liquid 1000
		4000	3250	2500	1750	1000		
Properties of heptanes plus								
Specific Gravity	0.7976	0.7900	0.7776	0.7652	0.7523	0.7381		0.8181
Molecular Weight	152	145	136	128	120	112		171
GPM content of produced well stream (Gal./MSCF)								
Propane	1.421	1.407	1.396	1.387	1.384	1.485		
Iso-Butane	0.664	0.644	0.631	0.624	0.631	0.674		
N-Butane	0.524	0.511	0.502	0.498	0.495	0.518		
Iso-Pentane	0.437	0.418	0.407	0.396	0.396	0.411		
N-Pentane	0.192	0.177	0.174	0.170	0.177	0.177		
Hexanes	0.730	0.667	0.633	0.608	0.587	0.600		
Heptanes plus	3.904	2.992	2.018	1.508	1.232	1.103		
Totals	7.872	6.816	5.761	5.191	4.902	4.968		
Deviation factor "Z" of well stream produced	0.942	0.910	0.876	0.867	0.882	0.911		
Calculated viscosity of well stream produced (CP)	0.0374	0.0317	0.0251	0.0206	0.0174	0.0152		
Well stream produced cumulated percentage	0.00	5.98	20.01	36.71	54.73	72.95		

TABLE 1.8 Retrograde Condensation During Gas Depletion at 285°F [11]

| Pressure (PSIA) | Reservoir Liquid | |
	(BBL./MMSCF of Dew Point Fluid)	(Volume* Percent)
4420 D.P. 285°F	0.00	0.00
4000	10.58	1.30
3250	81.47	9.98
2500	101.57	12.46
1750	103.69	12.70
1000	97.34	11.92

*Percent of reservoir hydrocarbon pore space @ dew point.

and adjust it to separator or flash conditions by

$$B_o = B_{od} \frac{B_{ofb}}{B_{odb}}$$

This calculation is illustrated in Figure 1.22.

To perform material-balance calculations, we must also have the separator and stock-tank gas in solution as a function of reservoir pressure. These values are expressed as standard cubic feet per barrel and usually are designated R_{sf}. The separator test gives us this value at the bubble point, R_{sfb}. As pressure declines in the reservoir, gas is evolved from solution. The amount of gas remaining in solution in the oil is then somewhat less. The differential vaporization tells us how much gas was evolved from the oil in the reservoir: $(R_{sdb} - R_{sd})$, where R_{sdb} is the amount of gas in solution at the bubble point as measured by differential vaporization at the reservoir temperature and R_{sd} is the gas in solution at subsequent pressures.

The units of R_{sdb} and R_{sd} are standard cubic feet per barrel of residual oil. Because we must have the gas in solution in terms of standard cubic feet per barrel of stock-tank oil, this term must be converted to a stock-tank basis. If we divide $(R_{sdb} - R_{sd})$ by B_{odb}, we have the gas evolved in terms of standard cubic feet per barrel of bubble point oil. If we then multiply by B_{ofb}, we will have the gas evolved in terms of standard cubic feet per barrel of stock-tank oil. This expression now is $(R_{sdb} - R_{sd})(B_{ofb}/B_{odb})$. The gas remaining in solution then is $R_s = R_{sfb} - (R_{sdb} - R_{sd})(B_{ofb}/B_{odb})$ standard cubic feet per stock-tank barrel. For every pressure studied during the differential liberation, R_s may be calculated from this equation. This calculation is illustrated in Figure 1.23.

It is a fairly common practice to use differential vaporization data for material-balance calculations. Values of B_{od} and R_{sd} are almost always higher than the corresponding values from separator tests; consequently, calculations of OIP and recoverable oil will usually be lower than is correct. The differential vaporization data should be converted to separator flash conditions before use in calculations.

TABLE 1.9 Calculated Cumulative Recovery During Depletion [11]

Cumulative Recovery Per MMSCF of Original Fluid	Initial in Place	Dew Point Pressure 4420	Reservoir Pressure — PSIA				
			4000	3250	2500	1750	1000
Well stream — MSCF	1000.00	0.0	59.80	200.10	367.10	547.30	729.50
Normal temperature separation stock tank liquid — barrels							
Cumulative produced.	0.0	0.0	7.06	19.79	30.80	41.53	50.84
Remaining in vapor in res.	131.31	131.31	97.16	55.34	34.07	20.89	11.77
Remaining in liquid in res.	0.0	0.0	27.10	56.18	66.44	68.90	68.70
Primary sep. gas — MSCF	794.65	0.0	48.49	166.85	312.88	473.15	635.04
Second stage gas — MSCF	82.58	0.0	4.59	13.72	22.59	31.22	40.51
Stock-tank gas — MSCF	14.75	0.0	0.83	2.56	4.31	6.06	7.95
*Total "plant products" in primary separator gas — gallons***							
Propane plus	1822.09	0.0	111.98	392.31	754.17	1161.68	1587.72
Butanes plus	811.70	0.0	50.35	179.85	353.88	554.50	764.40
Pentanes plus	250.79	0.0	15.72	57.75	117.78	189.38	265.12
Total "plant products" in well stream — gallons							
Propane plus	7872.05	0.0	439.19	1321.39	2235.80	3144.93	4043.82
Butanes plus	6451.15	0.0	354.64	1040.24	1722.29	2381.70	3019.14
Pentanes plus	5263.11	0.0	284.57	809.73	1303.52	1760.39	2186.72

*Primary separator at 915 PSIA and 95°F., second stage separator @ 65 PSIA and 70°F., Stock-tank at 15 PSIA and 75°F.
**Recover assumes 100 percent plant efficiency.
All gas volumes calculated at 15.025 PSIA and 60°F and stock-tank liquid measured at 60°F.

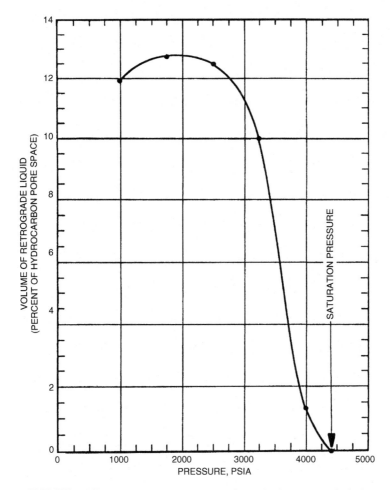

FIGURE 1.17 Retrograde condensation during depletion at 285°F [11].

1.2.3 Vapor–Liquid Equilibrium Calculations

The basic equilibrium calculations are the bubble point, dew point, and flash (or two-phase equilibrium). In the general flash calculation, the temperature and pressure are usually fixed and L/f is the dependent variable. All equilibrium calculations are based on the definition of the K value, such that

$$K_i = \frac{y_i}{x_i} = \frac{\text{concentration of "i" component in vapor phase}}{\text{concentration of "i" component in liquid phase}} \qquad (1.16)$$

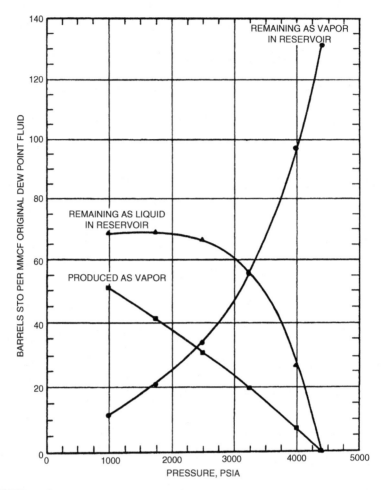

FIGURE 1.18 Stock-tank liquid production and retrograde condensation during constant volume production at 285°F [11].

TABLE 1.10   Calculated Wellstream Yields* [11]

Pressure, PSIA	4420 D.P.	4000	3250	2500	1750	1000
Gas–Liq. ratio**						
SCF 1st. Stg. gas						
BBL. S.T. Liq.	6052	7899	11,194	14,617	16,842	17,951
BBLS. S.T. Liq.						
MMSCF 1st Stg. gas	165.23	126.60	89.33	68.41	59.38	55.71
BBLS. S.T. Liq.						
MMSCF well stream Fld.	131.31	104.70	76.86	61.50	52.86	49.40

*Primary separator at 915 PSIA and 95°F., Second stage separator at 65 PSIA and 70°F, stock-tank at 15 PSIA and 70°F.
**All gas volumes calculated at 15.205 PSIA and 60°F and stock-tank liquid measured at 60°F.

TABLE 1.11   Differential Vaporization at 220°F [66]

Pressure PSIG	Solution Gas/Oil Ratio(1) R_{sd}	Relative Oil Volume(2) B_{od}	Relative Total Volume(3)	Oil Density gm/cc	Deviation Factor Z	Gas Formation Volume Factor(4)	Incremental Gas Gravity
2620	854	1.600	1.600	0.6562			
2350	763	1.554	1.665	0.6655	0.846	0.00685	0.825
2100	684	1.515	1.748	0.6731	0.851	0.00771	0.818
1850	612	1.479	1.859	0.6808	0.859	0.00882	0.797
1600	544	1.445	2.016	0.6889	0.872	0.01034	0.791
1350	479	1.412	2.244	0.6969	0.887	0.01245	0.794
1100	416	1.382	2.593	0.7044	0.903	0.01552	0.809
850	354	1.351	3.169	0.7121	0.922	0.02042	0.831
600	292	1.320	4.254	0.7198	0.941	0.02931	0.881
350	223	1.283	6.975	0.7291	0.965	0.05065	0.988
159	157	1.244	14.693	0.7382	0.984	0.10834	1.213
0	0	1.075		0.7892			2.039
		@ 60°F. = 1.000					

Gravity of residual oil = 35.1° API @ 60°F.

(1) Cubic feet of gas at 14.65 PSIA and 60°F per barrel of residual oil at 60°F.
(2) Barrels of oil at indicated pressure and temperature per barrel of residual oil at 60°F.
(3) Barrels of oil plus liberated gas at indicated pressure and temperature per barrel of residual oil at 60°F.
(4) Cubic feet of gas at indicated pressure and temperature per cubic foot at 14.65 PSIA and 60°F.

TABLE 1.12   Separator Test [8]

Separator Pressure (psig)	Temperature (°F)	GOR, R^*_{sfb}	Stock-Tank Oil Gravity (°API at 60°F)	FVF, B^{**}_{ofb}
50 to 0	75	737		
	75	41	40.5	1.481
		778		
100 to 0	75	676		
	75	92	40.7	1.474
		768		
200 to 0	75	602		
	75	178	40.4	1.483
		780		
300 to 0	75	549		
	75	246	40.1	1.495
		795		

*GOR in cubic feet of gas at 14.65 PSIA and 60°F per barrel of stock-tank oil at 60°F.
**FVF is barrels of saturated oil at 2620 PSIG and 220°F per barrel of stock-tank oil at 60°F.

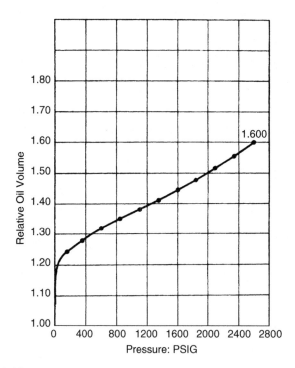

FIGURE 1.19   Adjustment of oil relative volume curve to separator conditions [6].

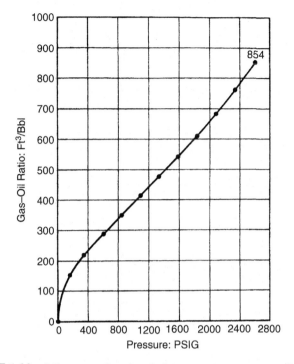

FIGURE 1.20   Adjustment of gas in solution curve to separator conditions [6].

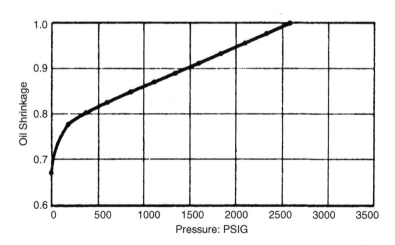

FIGURE 1.21   Oil-shrinkage curve [8].

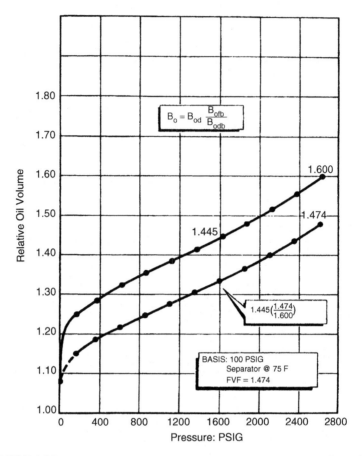

FIGURE 1.22   Adjustment of oil relative volume curve to separator conditions [6].

In bubble point calculations, x_i is known and either T or P is fixed. The vapor phase composition (y_i) and the P or T of the system are unknown.

$$y_i = K_i x_i \tag{1.17}$$

Several different values of the dependent variable are assumed. The correct value is the one that yields

$$\sum_{i=1}^{n} K_i x_i = \sum_{i=1}^{n} y_i = 1.0 \tag{1.18}$$

An additional requirement when using composition dependent K values is

$$\left| y_i^{m+1} - y_i^m \right| \leq \varepsilon, \quad i = 1 \text{ to } n \tag{1.19}$$

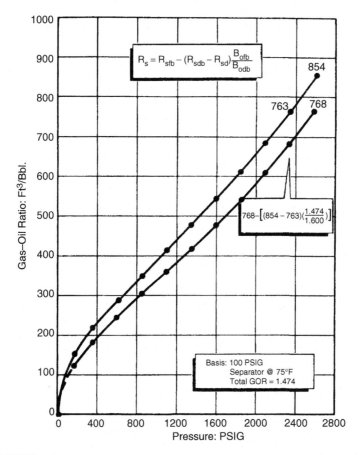

FIGURE 1.23 Adjustment of gas in solutions curve to separator conditions [6].

where the value of ε is arbitrarily small (10^{-4} to 10^{-6}). This requirement is a consequence of using composition dependent K values. If the composition y_i is not correct, the predicted K, values will not be correct. As a result, the composition of the vapor phase must be stabilized even though the correct value of the dependent variable has been determined. Iterations through the bubble point calculation must be continued until both Equation 1.18 and Equation 1.19 are satisfied. A logical diagram illustrating the basic bubble calculation is shown in Figure 1.24.

How can we assume initial vapor phase composition? One approach that has been successful is to assume that the mole fraction of the lowest boiling component in the systems is equal to unity with the remaining component mole fractions set to 10^{-6}. Another approach is to get the K value from GPSA [2]. The vapor composition is adjusted after each interaction.

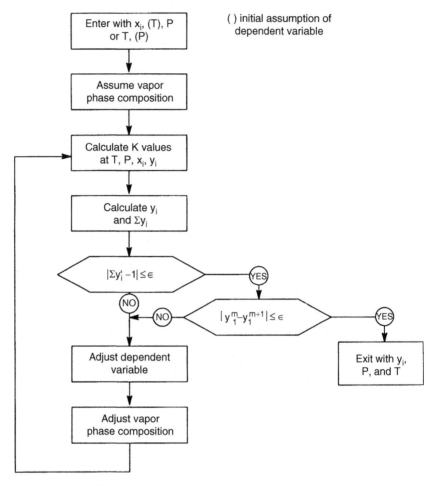

FIGURE 1.24 Block diagram for bubble point calculations.

Dew point calculations are the opposite of bubble point calculations: y_i is known and x_i and T or P are to be calculated. The specific equation used in the dew point calculation is

$$x_i = \frac{y_i}{K_i} \tag{1.20}$$

As in the bubble point calculation, several different values of the dependent variable are assumed:

$$\sum_{i=1}^{n} \frac{y_i}{K_i} = \sum_{i=1}^{n} x_i = 1.0 \tag{1.21}$$

subject to the condition that

$$\left|x_i^{m+1} - x_i^m\right| \le \varepsilon, \quad i = 1 \text{ to } n \tag{1.22}$$

A logical diagram for dew point calculation is shown in Figure 1.25. In this case x_i is initially unknown and must be assumed. A procedure is to assume that the mole fraction of the highest boiling component in the system is equal to unity. The remaining component mole fractions are set to 10^{-6}. Liquid phase compositions are adjusted by linear combinations of the assumed and calculated value during each iteration. The convergence of algorithms for T and P dependent calculations are given in Figure 1.26 and Figure 1.27.

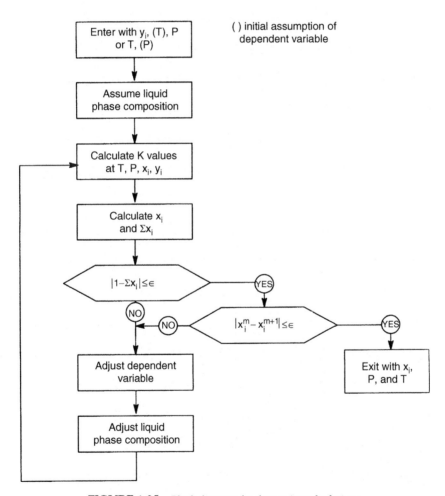

FIGURE 1.25 Block diagram for dew point calculations.

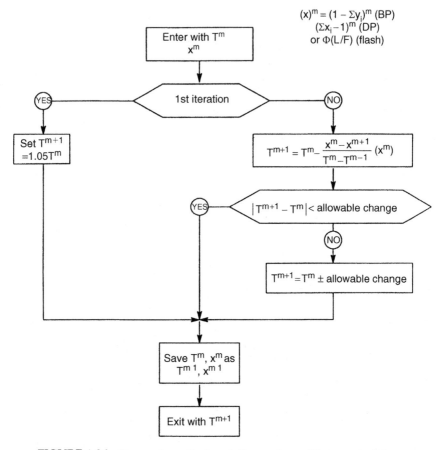

FIGURE 1.26   Temperature adjustment diagram for equilibrium calculations.

The purpose of flash calculations is to predict the composition and amount of the coexisting vapor and liquid phases at a fixed temperature and pressure.

According to Equations 1.15 and 1.16

$$\phi\left(\frac{L}{P}\right)^m = \sum_{i=1}^{n} \frac{z_i(1-K_i)}{L/F(1-K_i)+K_i} = 0.0 \tag{1.23a}$$

or

$$\phi\left(\frac{L}{P}\right)^m = 1.0 \tag{1.23b}$$

This equation is applicable to a wide range of L/F conditions.

In the basic flash calculation, T, P and the overall composition (z_i) are fixed. The unknown variables are x_i, y_i, and L/F. A convergence algorithm

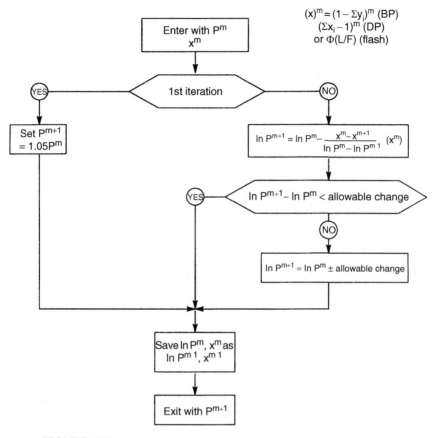

FIGURE 1.27 Pressure adjustment diagram for equilibrium calculations.

that can be used with Equation 1.24 is

$$\left(\frac{L}{F}\right)^{m+1} = \left(\frac{L}{F}\right)^m - \frac{\phi\left(\frac{L}{F}\right)^m}{\phi\left(\frac{L}{F}\right)^{m-1}} \qquad (1.24a)$$

where

$$\phi\left(\frac{L}{P}\right)^m = -\sum_{i=1}^{n} \frac{z_i(1-K_i)^2}{L/F(1-K_i)+K_i^2} \qquad (1.24b)$$

This convergence algorithm is very reliable provided the values of $(L/F)^{m+1}$ are constrained to be valid by material balance considerations:

$$0.0 < (L/F)^{m+1} \leq 1.0 \qquad (1.25)$$

The classical bubble point/dew point checks

$$\sum K_i z_i > 1.0 \quad \text{and} \quad \sum z_i/K_i > 1.0 \tag{1.26}$$

(to assure that the mixture is in two-phase region) cannot be conveniently used in most computer equation-of-state–based flash calculations because the K values for a given system are not known until the final solution has been reached. Consequently, the flash calculation (and its convergence algorithm) must be capable of performing "flash calculation" on single-phase systems (subcooled liquids, superheated vapors, and dense gas systems) as well as reliably predicting the amount of vapor and liquid present in a two-phase system. When the above flash equation/convergence algorithm is used on single-phase systems, the final predicted value of L/F will usually be outside the interval described by Equation 1.25 unless the material balance constraint is enforced. Should a value of $(L/F)^{m+1}$ outside the limits defined by Equation 1.25 be detected in an interaction, we recommend that the value of L/F predicted by Equation 1.24a be replaced by the appropriate value described by the following equations:

$$\text{if } (L/F)_{m+1} < 0.0, \quad (L/F)^{m+1} = (L/F)^m/2.0$$

or

$$\text{if } (L/F)^{m+1} > 1.0, \quad (L/F)^{m+1} = [1 + (L/F)^m]/2.0$$

This procedure eliminates most of the problems associated with flash calculations in single-phase regions and yields excellent results in relatively few iterations inside the two-phase region. Some problems still occur when attempting flash calculations in the dense gas region.

Initial estimates of the phase composition must be made to initiate the flash calculation. Several procedures are available. It was found that a combination of the bubble point/dew point initial phase estimation procedures works quite well [12]. Set the vapor phase mole fraction of the highest component in the system at 1.0 and the liquid phase mole fraction of the heaviest component in the system at 1.0. All other mole fractions are set to 10^{-6}. This procedure is believed to be superior to the technique of basing the initial assumption of the phase composition on some noncomposition-dependent K value estimation procedure, particularly when a wide range of temperatures, pressures, component types, composition ranges, etc., is to be considered.

The estimated vapor and liquid phase compositions must be compared with the calculated phase compositions. Equations 1.19 and 1.23 describe this checking procedure. If restraints described by these equations for any component (in either phase) are not satisfied, the calculations must be repeated even though an acceptable value for L/F has been determined. Some feel that this detailed checking procedure is unnecessary. It probably is unnecessary for most problems involving moderate

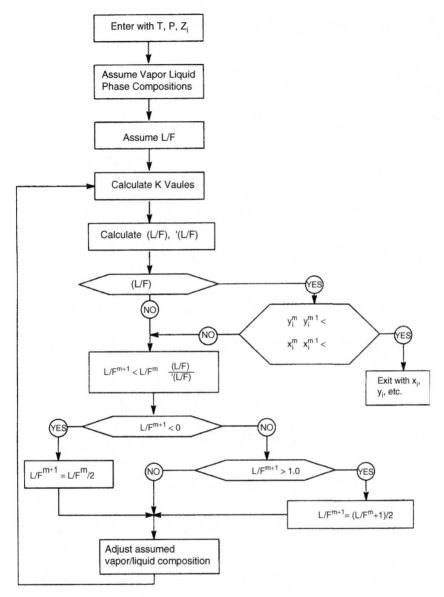

FIGURE 1.28 Block diagram for two-phase equilibrium calculations.

temperature–pressure-composition conditions. However, at extreme condition of temperature, pressure, and composition (low-temperatures, high pressure, high-acid-gas compositions) failure to perform these composition checks will lead to results that are completely incorrect (poor estimates of the phase compositions and incorrect L/F ratios). Unfortunately, the boundary changes in temperature, pressure, or composition can completely alter the difficulty of a given problem. Consequently, careful application of these checks in all calculations is strongly recommended since one can never be sure that a particular problem will not fall into the area of extreme conditions.

A logic diagram illustrating the basic flash calculation is shown in Figure 1.28. All the necessary features described earlier are embodied in this diagram.

Flash calculations at fixed L/F and temperature or pressure are frequently necessary. In these calculations, the dependent variable becomes pressure or temperature, and the flash calculation becomes similar in principle to a bubble point or dew point calculation. The flash calculation equation described earlier, Equation 1.23b, can be coupled with the temperature or pressure adjusting algorithms described forth bubble point/dew point calculations to perform these calculations. Initial estimates of the vapor- or liquid-phase compositions must be made, and the approach described in the flash at fixed temperature–pressure conditions can be used quite effectively. The logic diagram for this type of calculation can be deduced from earlier diagrams.

1.2.4 Predicting the Properties of Hexane Plus (C_{6+}) Fractions

Physical properties of light hydrocarbons are given in Table 1.1. In naturally occurring gas and oil, C_{6+} is unknown and makes a problem. Since the C_{6+} is a combination of paraffins (P), napthenes (N), and aromatics (A) of varying molecular mass (M), these fractions must be defined or characterized in some way. Changing the characterization of C_{6+} fractions present in even small amounts (at 1.0% mole level) can have significant effect on the predicted phase behavior of a hydrocarbon system. The dew point of the gas is heavily dependent upon the heaviest components in the mixture.

The SRK (Equation 1.11) and the PR (Equation 1.13) require the smallest number of parameters of any of the equations of state. They require the critical temperature, the critical pressure, and the acentric factor. There are many different approaches that can be utilized to predict these parameters for C_{6+} fractions or other mixtures of undefined components.

Some minimum of information must be available on the C_{6+} fraction, usually it is specific gravity (S), average boiling point (T_b), and molecular mass (M) of the fraction.

The following equation is used [14] to estimate the molecular mass (M) of petroleum fractions

$$M = 2.0438 \times 10^2 \exp(0.00218T) \exp(-3.07S) T^{0.118} S^{1.88} \qquad (1.27)$$

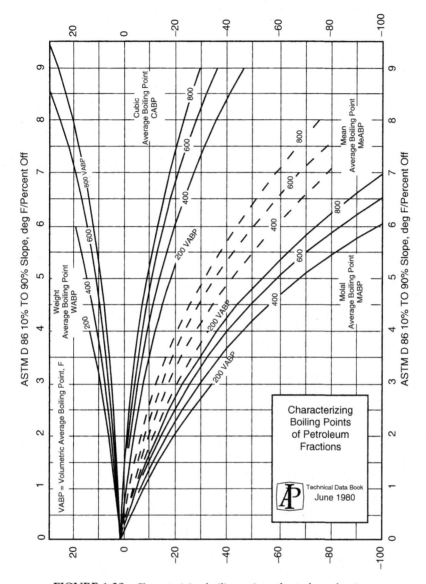

FIGURE 1.29   Characterizing boiling points of petroleum fractions.

where T = mean average boiling point of a petroleum
fraction, °R (from ASTM D86 test, Figure 1.29)
S = specific gravity, 60°F/60°F

The following equation is to be used to calculate the initial temperature (T_c) of pure hydrocarbons; it is applicable for all families of hydrocarbons:

$$\log T_c = A + B \log S + C \log T_b \qquad (1.28)$$

$T_c(°R)$; T_b and S given; A, B, and C as below:

Type Compound	A	B	C
Paraffin	1.47115	0.43684	0.56224
Napthene	0.70612	-0.07165	0.81196
Olefin	1.18325	0.27749	0.65563
Acetylene	0.79782	0.30381	0.79987
Diolefin	0.14890	-0.39618	0.99481
Aromatic	1.14144	0.22732	0.66929

For petroleum fractions, physical properties can be predicted more accurately if the fraction of paraffins (P), naphthenes (N), and aromatics (A) are known. If θ is a physical property to be predicted and the molecular type fractions are known, a pseudocompound, i.e., a compound having the same boiling point and specific gravity as the fraction, for each molecular type can be defined. These properties can be combined by

$$\theta = P\theta_P + N\theta_N + A\theta_A \qquad (1.29)$$

where $P+N+A=1.0$, are the sum of paraffins, napthenes, and aromatics present in the fraction.

The basic equation for any value of property θ is [15]

$$\theta = aT_b^b S^c \qquad (1.30)$$

Correlation constants are given in Table 1.13.

Next, correlation of critical properties and acentric factor (ω) of hydrocarbons and derivatives is developed in terms of M, T_b, and S. Since molecular mass is readily determined by experiments, it is introduced as a correlating variable to obtain more general results. The specific gravity is the ratio of the density of the liquid at 20°C to that of water at 4°C.

The critical properties and acentric factor of C_1 to C_{20} n-alkanes are correlated with M by the following equation:

$$\theta_A = C_1 + C_2 M + C_3 M^2 + C_4 M^3 + C_5/M \qquad (1.31)$$

TABLE 1.13 Correlation Constants for Equation 1.30

θ	a	b	c
M (Molecular mass)	4.5673×10^{-5}	2.1962	-1.0164
T_c (Critical Temperature °R)	24.2787	0.58848	0.3596
P_c (Critical Pressure, psia)	3.12281×10^9	-2.3125	2.3201
V_c^m (Molar critical volume ft.^{3}/lb mole)	7.0434×10^{-7}	2.3829	-1.683
V_c (Critical Volume ft.^{3}/lb)	7.5214×10^{-3}	0.2896	-0.7666
V (Liquid molar volume at 20°C and 1 atm (cm^3/g mole))	7.6211×10^{-5}	2.1262	-1.8688
ρ (Liquid density g/cc)	0.982554	0.002016	1.0055

TABLE 1.14 Coefficients for θ_A in Equation 1.31

θ_A	C_1	C_2	C_3	C_4	C_5
T_c	2.72697×10^2	3.91999	-1.17706×10^{-2}	1.48679×10^{-5}	-2.27789×10^3
$\ln P_c$	1.77645	-1.01820×10^{-2}	2.51106×10^{-5}	-3.73775×10^{-8}	3.50737
V_c	1.54465×10	4.04941	1.73999×10^{-4}	1.05086×10^{-6}	2.99391×10^2
ωT_c	-1.56752×10	1.22751	9.96848×10^{-3}	-2.04742×10^{-5}	-6.90883×10
S	6.64050×10^{-1}	1.48130×10^{-3}	-5.07021×10^{-6}	6.21414×10^{-9}	-8.45218
T_b	1.33832×10^2	3.11349	-7.08978×10^{-3}	7.69085×10^{-6}	-1.12731×10^3

T_c and T_b are in K, P_c in MPa, V_e in cm^3/mol.

where θ_A represents T_c, L_nP_c, V_c, or ωT_c of a n-alkane. The coefficients C_1 to C_5 are reported in Table 1.14 for each property. It was additionally correlated S and T_b of the n-alkanes by Equation 1.31, and the coefficients are included in Table 1.14. The correlated S_A and T_{bA} of the n-alkanes will be required in the perturbation equations to follow as independent variables.

The average absolute deviations (AAD) of the correlations from the American Petroleum Institute project 44 table values are 0.15% for T_c, 1.0% for P_c (excluding methane), 0.8% for V_c, 1.2% for ω, 0.11% for T_b, and 0.07% for S. The specific gravity correlation applies only to C_5–C_{16}, which are the only n-alkanes that are liquids at 20°C.

Properties of the general hydrocarbons and derivatives are correlated as perturbations of those n-alkanes according to the equation

$$\theta = \theta_A + A_1\Delta S + A_2\Delta T_b + A_3(\Delta S)^2 + A_4(\Delta S)(\Delta T_b) + A_5(\Delta T_b)^2$$
$$+ A_6(\Delta S)^3 + A_7(\Delta S)^2(\Delta T_b) + A_8(\Delta S)(\Delta T_b)^2 + A_g(\Delta T_b)^3 \quad (1.32a)$$

with

$$\Delta S = S - S_A \quad (1.32b)$$
$$\Delta T_b = T_b - T_{bA} \quad (1.32c)$$

where S_A and T_{bA} are the gravity and boiling point of the hypothetical n-alkane of the M of the substance of interest and are given by Equation 1.31. The coefficients A_i in Equation 1.32a are given by

$$A_i = a_i + b_iM \quad (1.32d)$$

Table 1.15 presents the coefficients a_i and b_i that have been determined by fitting Equation 1.32d to the properties of a large number of hydrocarbons and derivatives.

State-of-the-art Equation 1.32a gives the best results.

The five general categories of experimental data availability for C_{6+} fractions are:

1. *The specific or API gravity of the C_{6+} fraction.* The molecular mass may have been determined experimentally or estimated from some correlation of specific gravity and molecular mass.

TABLE 1.15   Coefficients for Equation 1.32d

	θ			
	T_c	lnP_c	V_c	$\omega\,T_c$
a_1	1.58025×10^3	9.71572	-1.18812×10^3	-1.16044×10^3
a_2	-5.68509	-3.32004×10^{-2}	-1.18745	3.48210
a_3	-1.21659×10^4	-8.60375×10	7.36085×10^3	2.78317×10^4
a_4	7.50653×10	5.50118×10^{-1}	6.83380×10	-2.05257×10^2
a_5	-9.66385×10^{-2}	-9.00036×10^{-4}	-2.12107×10^{-1}	4.55767×10^{-1}
a_6	2.17112×10^4	1.85927×10^2	-4.84696×10^3	-7.13722×10^4
a_7	-1.57999×10^2	-1.51115	-4.12064×10^2	5.08888×10^2
a_8	3.60522×10^{-1}	4.32808×10^{-3}	2.02114	-6.10273×10^{-1}
a_9	-2.75762×10^{-4}	-3.81526×10^{-6}	-2.48529×10^{-3}	-1.68712×10^{-3}
b_1	-1.18432×10	-7.50370×10^{-2}	1.17177×10	1.89761
b_2	5.77384×10^{-2}	3.15717×10^{-4}	-3.48927×10^{-2}	2.41662×10^{-2}
b_3	1.10697×10^2	8.42854×10^{-1}	-1.34146×10^2	-2.67462×10^2
b_4	-6.58450×10^{-1}	-5.21464×10^{-3}	5.63667×10^{-2}	2.06071
b_5	7.82310×10^{-4}	7.87325×10^{-6}	9.52631×10^{-4}	-5.22105×10^{-3}
b_6	-2.04245×10^2	-1.85430	1.80586×10^2	7.66070×10^2
b_7	1.32064	1.36051×10^{-2}	2.56478	-5.75141
b_8	-2.27593×10^{-3}	-3.23929×10^{-5}	-1.74431×10^{-2}	8.66667×10^{-3}
b_9	8.74295×10^{-7}	2.18899×10^{-8}	2.50717×10^{-5}	1.75189×10^{-5}

2. *Chromatographic analysis.* The C_{6+} fraction has been analyzed by gas–liquid chromatography. These results may be reported as a series of equivalent n-paraffins up to as high as nC_{30} or as a true boiling point analysis. The specific gravity and/or molecular mass of the fraction may or may not be reported.

3. *ASTM-D158 (or equivalent) analysis.* This analysis is equivalent to a nonrefluxed single-stage batch distillation. Usually the boiling point temperature at seven different points (START, 10%, 30%, 50%, 70%, 90%, END) will be recorded. The points correspond to the volume fraction of the C_{6+} distilled into the receiver vessel. The specific gravity and molecular mass of the total C_{6+} fraction are normally also measured.

4. *A partial TBP analysis.* A true boiling point (TBP) distillation has been performed on the C_{6+} fraction. The TBP distillation is a batch distillation similar to an ASTM distillation but the distillation apparatus contains several trays (usually 10 or more or the equivalent amount of packing) and a high reflux ratio is used. The TBP gives a sharper separation between the subfractions than an ASTM distillation. Normally, at least five temperatures are reported as a function of liquid volume percent distilled over. Frequently, more than 20 temperatures will be reported.

The specific gravity and molecular mass of the total fraction are usually reported.

5. *A complete TBP analysis.* A true boiling point distillation has been performed on the total C_{6+} fraction. The specific gravity and molecular mass have been measured for each of the reported distillate subfractions. Between 5 and 50 temperatures and subfraction properties will be reported.

Table 1.15 shows typical information as it may be reported for each of the five categories of C_{6+} characterization. The complete TBP analysis is believed to be the best form of C_{6+} analysis to be used with today's thermodynamic property prediction procedures. Consequently, it is recommended that all noncomplete TBP analyses be converted to this form. This section deals with these conversion techniques. These techniques are based on empirical correlations and, in some cases, experience and judgment. There is also one basic constraint that must be used in these conversion techniques — that is, maintenance of volume-mass-molar relationships in the C_{6+} fraction along with consistency in the composition of the total stream. One cannot capriciously change the molecular mass or specific gravity of the total C_{6+} fraction without simultaneously adjusting the reported composition. All of the procedures reported here strive to maintain consistency of the specific gravity, molecular mass and, when possible, the boiling point (s) of the total C_{6+} fraction.

The various procedures for converting noncomplete TBP analyses to complete TBP analyses are illustrated in the following section. A common sample problem is used to illustrate the basic conversion procedure. In addition, the results of several equilibrium calculations are reported for each type of characterization. The gas composition, true boiling point date, gravity, and molecular weight measurements for the C_{7+} fraction are shown in Table 1.16. Though the particular system chosen shown C_{7+} as a basis for the heavy and characterization C_{6+} will be used. There are several isomers of hexane, as well as other materials, that can appear in the C_{6+} subfraction. The molecular mass tabulated for the fractions in Table 1.16 makes them appear to be normal paraffins. This, however, is not true and a complete TBP analysis was made on the C_{7+} fraction.

Calculations made based on the different C_{7+} characterizations are compared with experimental values, Table 1.17 and Figure 1.30. The complete TBP characterization provides the best predictions of the phase behavior and the liquid formation, though there is only a little difference between the full TBP and the partial TBP results. The lumped specific gravity-molecular mass characterization and the lumped n-paraffin characterization give the poorest predictions. All of the characterizations in Table 1.18 are in better agreement with experimental values than one would normally expect.

TABLE 1.16 Sample Analysis for Five Categories of C_{6+} Analysis [12]

Categories

1 Specific Gravity		2* Molar Chromatographic		3 ASTM		4 Partial TBP		5 Complete TBP			
Information	Values Reported	Information Reported	Values mol%	Information Reported LV%	Values Reported T,°F	Information Reported LV%	Values Reported T,°F	Information Reported LV%	T,°F	sp gr	mol wt
specific gravity	0.7268	C_6	0.335	ST	258	ST	155	ST	155		
(°API)	(63.2)	C_7	0.327	10	247	5	190	17.52	238	0.745	100
molecular mass	104	C_8	0.341	30	283	10	212	33.12	280	0.753	114
		C_9	0.268	50	331	20	246	·	·	·	·
		C_{10}	0.166	70	403	30	270	·	·	·	·
				90	500	40	304	·	·	·	·
		C_{33}	0.004	GP	596	50	348	·	·	·	·
		C_{34}	0.004			60	380	·	·	·	·
		C_{35+}	0.006	specific gravity° API	0.7867	70	431	·	·	·	·
		specific gravity	?		48.37	80	481	·	·	·	·
						90	538	99.73	·	·	·
		molecular mass	?	molecular mass	141.26	95	583	EP	698	0.878	310
						EP	700				
						Specific gravity	0.7867				
						molecular mass	141.26				

*Chromatographic TBP will be similar to ASTM or partial TBP

TABLE 1.17 Experimental Data for Illustrative Calculations [12]

Component	mol %	BPT°F	sp gr	mol wt
C_1	91.35			
C_2	4.03			
C_3	1.53			
iC_4	0.39			
nC_4	0.43			
IC_5	0.15			
nC_5	0.19			
C_6	0.39			
Fraction 7	0.361	209	0.745	100
8	0.285	258	0.753	114
9	0.222	303	0.773	128
10	0.158	345	0.779	142
11	0.121	384	0.793	156
12	0.097	421	0.804	170
13	0.083	456	0.816	184
14	0.069	488	0.836	198
15	0.050	519	0.840	212
16	0.034	548	0.839	226
17	0.023	576	0.835	240
18	0.015	603	0.850	254
19	0.010	628	0.865	268
20	0.006	653	0.873	282
21	0.004	676	0.876	296
22	0.002	698	0.878	310

mol % $C_{7+} = 1.540$

mol wt $C_{7+} = 141.26$

sp gr $C_{7+} = 0.7867$ (43.35° API)

Phase Behavior Data

dew point at 201°F–3837 psia

Liquid Formation at 201°F

Pressure psia	Bbl/MMSCF	Specific Gravity Liquid
2915	9.07	0.6565
2515	12.44	0.6536
2015	15.56	0.6538
1515	16.98	0.6753
1015	16.94	0.7160
515	15.08	0.7209

TABLE 1.18 Effect of C_{7+} Characterization on Predicted Sample Problem Phase Behavior [12]

	Experiment Values	C_{7+} from Table 1.16	Partial TBP	Fractions as n-paraffins	C_{7+} Predicted Values for Characterization Chromatographic		
					Fraction from ASTM dest. curve	Lumped C_{7+}	C_{7+} as nC_{10}
Dew Point at 201°F, psig	3822	3824	3800	3583	3553	3150	2877
Amount of liquid at BBL/MMSCF							
2900 psig	9.07	11.57	11.50	11.07	7.37	5.95	0.0
2500 psig	12.44	14.26	14.22	14.89	10.21	12.40	14.09
2000 psig	15.56	16.43	16.44	17.77	12.66	17.09	18.77
1500 psig	16.98	17.48	17.49	19.08	14.00	19.35	21.74
1000 psig	16.94	17.38	17.39	18.99	14.15	19.91	22.25
500 psig	15.08	15.56	15.59	16.98	12.46	18.11	20.32

*Based on 10 equal LV% fractions

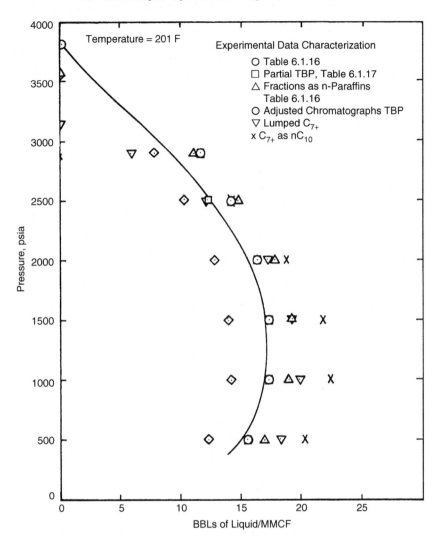

FIGURE 1.30   Effect of C_{6+} Characterization on predicted liquid formation [12].

1.3 VAPOR-LIQUID EQUILIBRIUM BY EQUATION OF STATE

Prediction of a vapor-liquid mixture is more complicated than prediction of pure component VLE.

The following condition equations for mixture VLE can be derived from classical thermodynamics:

$$T^V = T^L \qquad (1.33)$$

$$P^V = P^L \tag{1.34}$$

$$f_i^V = f_i^L \tag{1.35}$$

What does "f_i" mean?

The equilibrium constants, or K values, are defined as the ratio of vapor and liquid compositions:

$$K_i = \frac{y_i}{x_i} \tag{1.36}$$

The Gibbs free energy is a property of particular importance because it can be related to the equilibrium state and at the same time can be expressed as a function of T and P:

$$dG_i = -SdT + VdP \tag{1.37}$$

If Equation 1.37 is applied to an ideal gas it becomes

$$dG_i = RT \, d(\ln P) \tag{1.38}$$

If we now define a property such that Equation 1.38 will apply for all gases under all conditions of temperature and pressure

$$dG_i = RT \, d(\ln f_i) \tag{1.39}$$

f_i is called the fugacity of component "i" and has units of pressure. If Equation 1.39 is integrated for an ideal gas:

$$f_i = cP \tag{1.40}$$

"c" is constant and for an ideal gas is equal to 1.0. For real gases the only condition under which the gas will behave ideally is at zero pressure. This can be expressed as

$$\lim_{p \to 0} \frac{f_i}{P} = 1.0 \tag{1.41}$$

The fugacity of a single component in a mixture is defined in a manner similar to Equation 1.36

$$dG = RT \, d(\ln f_i) \tag{1.42}$$

and by analogy to the Equation 1.39

$$\lim_{p \to 0} \frac{f_i}{x_i P} = 1.0 \tag{1.43}$$

The fugacity is sometimes referred to as a *corrected pressure*. A more valuable parameter for use in correlative procedures would be a variable with characteristics similar to fugacity, but which ranged over a much smaller range of numbers. It is the fugacity coefficient ϕ_i.

For a pure component

$$\phi_i = \frac{f_i}{P} \tag{1.44}$$

For a component in a mixture

$$\phi_i = \frac{f_i}{x_i P} \tag{1.45}$$

The fugacity coefficients are readily calculated form P–V–T data for both, pure component and mixture.

P–V–T of a fixed composition system can be developed in pressure explicit forms, i.e.,

$$P = \phi(V, T) \tag{1.46}$$

The basic equation of state may be transformed to a compressibility factor Z, but the basic expression given by Equation 1.44 still applies. Equilibrium K values are predicted from fugacity, which is related to Gibbs free energy. One of the basic definitions of fugacity gases K values is

$$K = \frac{y_i}{x_i} = \frac{\phi_i^L}{\phi_i^v} = \frac{f_i^L/(Px_i)}{f_i^v/(Py_i)} \tag{1.47}$$

The fugacity coefficient ϕ_i is related to the pressure, volume and temperature by

$$\ln \phi_i = -\frac{1}{RT} \int_V^\infty \left[\left(\frac{\partial P}{\partial n_i} \right)_{T,V,n_j} - \frac{RT}{V} \right] dV - \ln Z \tag{1.48}$$

Applying the SRK equation

$$\ln \phi_i = -\ln \left(Z - \frac{Pb}{RT} \right) + (Z - 1)B_i' - \frac{a}{bRT}(A_i' - B_i') \ln \left(1 + \frac{b}{V} \right) \tag{1.49}$$

or

$$\ln \phi_i = -\ln(Z - B) + (Z - 1)B_1' - \frac{A}{B}(A_i' - B_i') \ln \left(1 + \frac{B}{Z} \right) \tag{1.50}$$

where

$$B_i = \frac{b_i}{b} \tag{1.51}$$

$$A_i' = \frac{1}{a} \left[2a_i^{0.5} \sum_j^N x_j a_j^{0.5}(1 - k_{ij}) \right] \tag{1.52}$$

Notations are as for Equations 1.11 and 1.12. For values k_{ij} see Tables 1.19 and 1.20. The data in Table 1.18 was prepared for the Peng-Robinson equation, while the data from Table 1.20 was used for the Benedict-Webb-Rubin equation modified by Starling [16].

TABLE 1.19 Values of Interaction Parameters k_{ij} for Use in the Peng-Robinson Equation

($k_{ij} = 100$)

	Methane	Ethane	Propane	i-Butane	n-Butane	i-Pentane	n-Pentane	n-Hexane	n-Heptane	n-Octane	n-Nonane	n-Decane	Nitrogen	Carbon Dioxide	Hydrogen Sulfide	Toluene	Benzene	Cyclohexane	Water
Methane	0.0	0.0	0.0	0.0	0.0	0.0	0.0	0.0	0.0	0.0	0.0	0.0	3.6	10.0	8.5	4.0	4.0	3.5	50.0
Ethane		0.0	0.0	0.0	0.0	0.0	0.0	0.0	0.0	0.0	0.0	0.0	5.0	13.0	8.4	2.0	2.0	2.0	48.0
Propane			0.0	0.0	0.0	0.0	0.0	0.0	0.0	0.0	0.0	0.0	8.0	13.5	7.5	2.0	2.0	2.0	48.0
i-Butane				0.0	0.0	0.0	0.0	0.0	0.0	0.0	0.0	0.0	9.5	13.0	5.0	0.0	0.0	0.0	48.0
n-Butane					0.0	0.0	0.0	0.0	0.0	0.0	0.0	0.0	9.0	13.0	6.0	0.0	0.0	0.0	48.0
i-Pentane						0.0	0.0	0.0	0.0	0.0	0.0	0.0	9.5	5.0	6.0	0.0	0.0	0.0	48.0
n-Pentane							0.0	0.0	0.0	0.0	0.0	0.0	10.0	5.0	6.5	0.0	0.0	0.0	48.0
n-Hexane								0.0	0.0	0.0	0.0	0.0	10.0	5.0	6.0	0.0	0.0	0.0	48.0
n-Heptane									0.0	0.0	0.0	0.0	10.0	5.0	6.0	0.0	0.0	0.0	48.0
n-Octane										0.0	0.0	0.0	10.0	11.5	5.5	0.0	0.0	0.0	48.0
n-Nonane											0.0	0.0	10.0	11.0	5.0	0.0	0.0	0.0	48.0
n-Decane												0.0	10.0	11.0	4.5	1.0	1.0	1.0	0.0
Nitrogen													0.0	2.0	18.0	18.0	16.0	10.0	0.0
Carbon Dioxide														0.0	10.0	9.0	7.5	10.0	0.0
Hydrogen Sulfide															0.0	0.0	0.0	0.0	0.0
Toluene																0.0	0.0	0.0	0.0
Benzene																	0.0	0.0	0.0
Cyclohexane																		0.0	0.0
Water																			0.0

TABLE 1.20 Values of Interaction Parameters k_{ij} Proposed by Starling

	Methane	Ethylene	Ethane	Propylene	Propane	i-Butane	n-Butane	i-Pentane	n-Pentane	Hexane	Heptane	Octane	Nonane	Decane	Undecane	Nitrogen	Carbon dioxide	Hydrogen Sulfide
Methane	0.0	1.0	1.0	2.1	2.3	2.75	3.1	3.6	4.1	5.0	6.0	7.0	8.1	9.2	10.1	2.5	5.0	5.0
Ethylene		0.0	0.0	0.3	0.31	0.4	0.45	0.5	0.6	0.7	0.85	1.0	1.2	1.3	1.5	7.0	4.8	4.5
Ethane			0.0	0.3	0.31	0.4	0.45	0.5	0.6	0.7	0.85	1.0	1.2	1.3	1.5	7.0	4.8	4.5
Propylene				0.0	0.0	0.3	0.35	0.4	0.45	0.5	0.65	0.8	1.0	1.1	1.3	10.0	4.5	4.0
Propane					0.0	0.3	0.35	0.4	0.45	0.5	0.65	0.8	1.0	1.1	1.3	10.0	4.5	4.0
i-Butane						0.0	0.0	0.08	0.1	0.15	0.18	0.2	0.25	0.3	0.3	11.0	5.0	3.6
n-Butane							0.0	0.08	0.1	0.15	0.18	0.2	0.25	0.3	0.3	12.0	5.0	3.4
i-Pentane								0.0	0.0	0.0	0.0	0.0	0.0	0.0	0.0	13.4	5.0	2.8
n-Pentane									0.0	0.0	0.0	0.0	0.0	0.0	0.0	14.8	5.0	2.0
Hexane										0.0	0.0	0.0	0.0	0.0	0.0	17.2	5.0	0.0
Heptane											0.0	0.0	0.0	0.0	0.0	20.0	5.0	0.0
Octane												0.0	0.0	0.0	0.0	22.8	5.0	0.0
Nonane													0.0	0.0	0.0	26.4	5.0	0.0
Decane														0.0	0.0	29.4	5.0	0.0
Undecane															0.0	32.2	5.0	0.0
Nitrogen																0.0	0.0	0.0
Carbon dioxide																	0.0	3.5
Hydrogen sulfide																		0.0

Applying Peng-Robinson equation

$$\ln \phi_i = -\ln\left(Z - \frac{Pb}{RT}\right) + (Z-1)B_i' - \frac{a}{2^{1.5}bRT}$$

$$\times (A_i' - B_i') \ln\left[\frac{V + (2^{0.5}+1)b}{V - (2^{0.5}-1)b}\right] \tag{1.53a}$$

or

$$\ln \phi_i = -\ln(Z - B) + (Z-1)B_1' - \frac{A}{2^{1.5}B}$$

$$\times (A_i' - B_i') \ln\left[\frac{Z + (2^{0.5}+1)B}{Z - (2^{0.5}-1)B}\right] \tag{1.53b}$$

where

$$B_i = \frac{b_i}{b} \tag{1.54}$$

$$A_i' = \frac{1}{a}\left[2a_i^{0.5}\sum_j^N x_j a_j^{0.5}(1 - k_{ij})\right] \tag{1.55}$$

Notations are as in Equation 1.13 and Equation 1.14.

The objectives of any equation-of-state solution method are the reliable and accurate prediction of the volumetric properties of the fluid mixture under consideration. The overall solution procedure is as follows:

- fix the total composition, temperature and pressure
- calculate constant for the equation of state
- solve equation for the volumetric property (specific volume, density or compressibility factor)

When pressure and temperature fall to a two-phase region, the equation must be solved twice, separately for vapor and liquid. The composition of each phase will be different, so the equation of state constants will have to be evaluated for both the liquid and the vapor phases. Both SRK and PR are cubic equations, so the solution always gives three roots, as shown in Figure 1.31. However, the P_r–V_r relationship at a given T_r is discontinuous at $V_r = b_1'$, $V_r = b_2'$, and $V_r = b_3'$. We are interested in only $V_r > b_1'$, which in case the SRK equation is equal 0.08664 and 0.077796 for the PR equation. For $V_r > b_1'$ and $T_r > 1.0$, there is only one value of the compressibility factor that will satisfy the equation of state. For $V_r > b_1'$ and $T_r < 1.0$, we will get three values of Z. The largest Z of the vapor Z's is chosen for the vapor and the smallest amount the liquid Z's is chosen for the

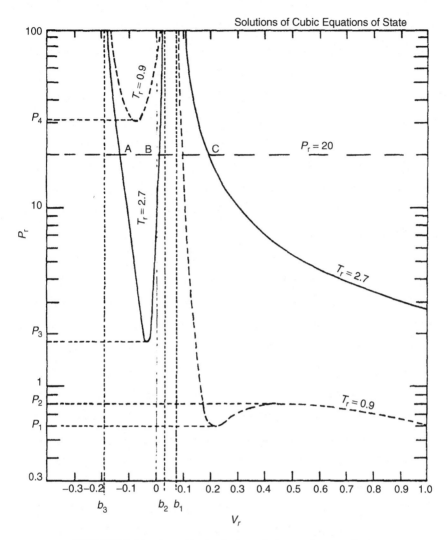

FIGURE 1.31 P_r vs. V_r plot for Peng-Robinson Equation [17].

liquid. However, in an earlier stage of the iterative VLE calculations, it is not uncommon to encounter a single root, mainly because of incorrect compositions.

A logic diagram for a trial-and-error solution procedure for cubic equations of state is given in Figure 1.32. This diagram shows a traditional Newton-Raphson approach with an interval halving limiting procedure superimposed on it. For purposes of the procedure for locating the boundaries of the liquid- and vapor-phase compressibility factors in the

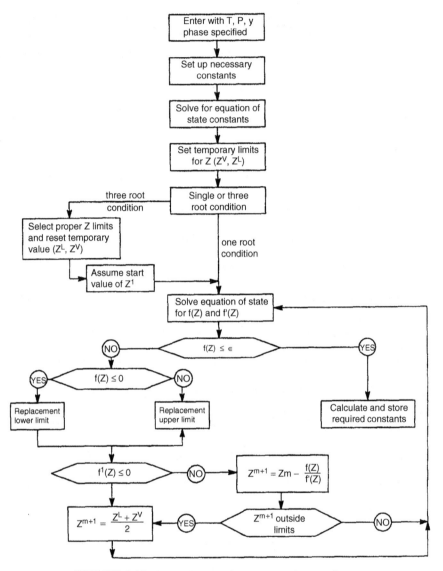

FIGURE 1.32   Logic diagram for equation-of-state solution.

two-phase region discussion, assume that the equation of state is given in the form

$$Z^3 + \alpha Z^2 + \beta z + \gamma = 0 \qquad (1.56)$$

where γ and β are arbitrary constants.

From Figure 1.33a–d in the two-phase region the equation of state will have a maximum and a minimum or two points at which the slope of the

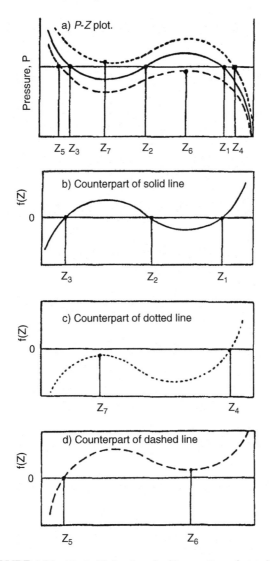

FIGURE 1.33   Typical behavior of cubic equation-of-state [17].

equation is zero. The value of the compressibility factor at the maximum defines the largest possible value for the liquid-phase compressibility factor. If Equation 1.56 is differentiated and set equal to zero the following equation results:

$$Z^2 + \frac{2\alpha Z}{3} + \frac{\beta}{3} = 0 \tag{1.57}$$

Solving the above equation for the values of Z at the maximum and minimum gives

$$Z_{1,2} = \frac{-2\frac{\alpha}{3} \pm 4\frac{\alpha^2}{9} - 4\frac{\beta}{3}}{2}$$ (1.58)

If the algebraic expression under the square root sign is negative or zero, only one real value of the compressibility factor will satisfy the equation of state (Figure 1.33c or d). If, however, the value of the expression under the radical is positive, three real roots exist and limits for the vapor and liquid phase compressibility factors can be determined from Equation 1.58. The solutions of Equation 1.58 represent the value of Z at the maximum and minimum points of Figure 1.33b. The value of the maximum will represent the largest possible value for the liquid compressibility factor and the value at the minimum represents the smallest possible value of the vapor compressibility factor. These limits can then be used with arbitrary values for the other limit to assure that the root obtained is the valid one. The limits thus set up are adjusted at the end of each iteration to narrow the interval of search.

References

[1] Katz, D. L., et al., *Handbook of Natural Gas Engineering*, McGraw-Hill Book Co., New York, 1959.
[2] *Engineering Data Book*, Vols. 1, 2, 11th Edition, GPSA, Tulsa, Oklahoma, 1998.
[3] SPE Reprint No.13, Vol. 1., pp. 233–235.
[4] Soave, G., "Equilibrium Constants from a Modified Redlich-Kwong Equation of State," *"Chemical Engineering Science,"* 1972.
[5] Peng, O. Y., and Robinson, D. B., "A New Two Constant Equation of State," *Industry and Engineering Chemistry Fundamentals*, 1976.
[6] *The Phase Behavior of Hydrocarbon Reservoir Fluids*, course given at Core Laboratories Inc., Dallas, Texas.
[7] MacDonald, R. C., "Reservoir Simulation with Interface Mass Transfer," *Report No. UT-71-2, University of Texas, Austin, Texas., 1971.*
[8] Moses, P. L., "Engineering Applications of Phase Behavior of Crude Oil and Condensate Systems," *Journal of Petroleum Technology*, July 1986.
[9] API 811-08800 Standard RP 44: "Recommended Practice for Sampling Petroleum Reservoir Fluids."
[10] Amex, J. W., Bass, D. M., and Whiting, R. L., *Petroleum Reservoir Engineering*, McGraw-Hill Book Co., New York, 1960.
[11] Weatherly Laboratories, Inc., *Manual 1984*, Lafayette, Louisiana, 1984.
[12] Maddox, R. N., and Erbar, J. H., *Gas Conditioning and Processing*, Campbell CPS, Norman, Oklahoma, 1982.
[13] Edmister, W. C., and Lee, B. I., *Applied Hydrocarbon Thermodynamics*, Vol. 1, Gulf Publishing Co., Houston, Texas, 1984.
[14] Danbert, T. E., and Danner, R. P., Technical Data Conference and Exhibition of the SPE of AIME, New Orleans, Louisiana, October 3–6, 1976.

[15] Danbert, T. E., "Property Predictions," *Hydrocarbon Processing*, March 1980.

[16] Starling, E., *Fluid Thermodynamic Properties for Light Petroleum Systems*, Gulf Publishing Co., Houston, Texas, 1973.

[17] Asselineau, L., Bogdanic, G., and Vidal, J., *Fluid Phase Equilibrium*, 1979.

[18] Gunderson, T., *Computer and Chemical Engineering*, 1982.

Flow of Fluids

Fluid is defined as a single phase of gas or liquid or both. Each sort of flow results in a pressure drop. Three categories of fluid flow: vertical, inclined, and horizontal are shown in Figure 2.1. The engineer involved in petroleum production operations has one principal objective to move the

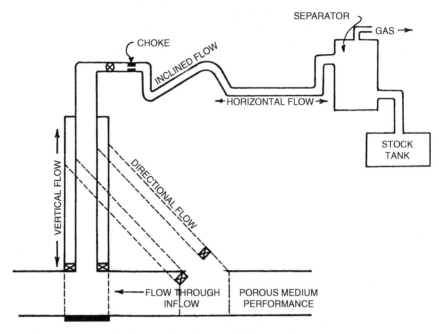

FIGURE 2.1 Overall production system [1].

fluid from some location in an underground reservoir to a pipeline that may be used to transport it or storage tank. Possible pressure losses in a complete production system and producing pressure profile are shown in Figures 2.2 and 2.3, respectively. On the way from reservoir to pipeline or storage tank, fluid is changing its temperature, pressure, and, consequently, composition of each phase. In case of a dry gas reservoir, a change in pressure and temperature does not create two-phase flow; also in case of black oil with a very small GOR, it could be assumed that two-phase flow does not occur.

Based on the law of conservation of energy, the total energy of a fluid at any particular point above datum plane is the sum of the deviation head, the pressure head, and velocity head as follows:

$$H = Z_{el} + \frac{144p}{\gamma} + \frac{v^2}{2g} \tag{2.1}$$

In reality, whenever fluid is moving there is friction loss (h_L). This loss describes the difference in total energy at two points in the system. Expressing the energy levels at point 1 versus point 2 then becomes

$$Z_{el1} + \frac{144p_1}{\gamma_1} + \frac{v_1^2}{2g} = Z_{el2} + \frac{144p_2}{\gamma_2} + \frac{v_2^2}{2g} + h_L \tag{2.2}$$

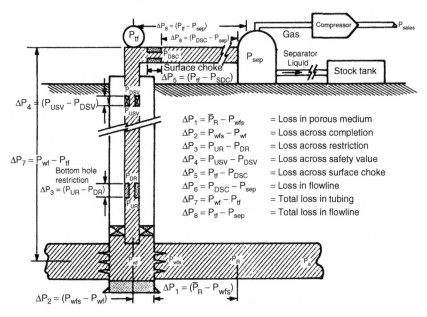

FIGURE 2.2 Possible pressure losses in complete system [1].

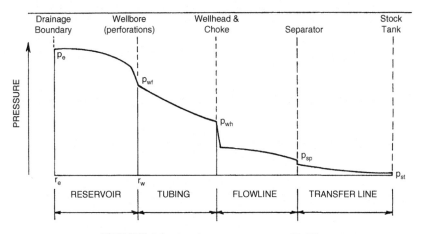

FIGURE 2.3 Production pressure profile [2].

All practical formulas for fluid flow are derived from the above,

where H = total energy of fluid

Z_{el} = pipeline vertical elevation rise (ft)

p_1, p_2 = inlet and outlet pressures (psia)

γ_1, γ_2 = inlet and outlet fluid specific weight

v_1, v_2 = inlet and outlet fluid velocity

g = acceleration due to gravity

h_L = loss of static pressure head due to fluid flow

Equation 2.2 can be written in differential form as

$$\frac{dp}{\gamma} + \frac{vdv}{g} + dL\sin\theta + dL_w = 0 \tag{2.3}$$

where $dL\sin\theta = dZ$, and dL_w refers to friction multiplying the equation by γ/dL to give

$$\frac{dp}{dL} = \frac{\gamma vdv}{gdL} + \gamma\sin\theta + \gamma\left(\frac{dL_w}{dL}\right)_f = 0 \tag{2.4}$$

Solving this equation for pressure gradient, and if we consider a pressure drop as being positive in the direction of flow,

$$\frac{dp}{dL} = \gamma\sin\theta + \frac{\gamma vdv}{gdL} + \left(\frac{dp}{dL}\right)_f \tag{2.5}$$

where

$$\left(\frac{dp}{dL}\right)_f = \frac{\gamma dL_w}{dL}$$

Equation 2.4 contains three terms that contribute to the total pressure gradient, i.e.:

1. pressure gradient due to elevation

$$\gamma\sin\theta = \left(\frac{dp}{dL}\right)_{el}$$

2. pressure gradient due to acceleration

$$\frac{\gamma vdv}{gdL} = \left(\frac{dp}{dL}\right)_{acc}$$

3. pressure gradient due to viscous forces (friction)

$$\frac{\gamma dL_w}{dL} = \left(\frac{dp}{dL}\right)_f$$

$$\frac{dp}{dL} = \left(\frac{dp}{dL}\right)_{el} + \left(\frac{dP}{dL}\right)_{acc} + \left(\frac{dp}{dL}\right)_f \tag{2.6}$$

The acceleration element is the smallest one and sometimes is neglected.

The total pressure at the bottom of the tubing is a function of flowrate and comprises three pressure elements:

1. wellhead pressure—back pressure exerted at the surface form choke and wellhead assembly
2. hydrostatic pressure—due to gravity and the elevation change between wellhead and the intake to the tubing
3. friction losses, which include irreversible pressure losses due to viscous drag and slippage

Figure 2.4 illustrates this situation for each single-phase and two-phase flow. Possible pressure losses in a complete system are shown in Figure 2.2. For a given flowrate, wellhead pressure, and tubing size, there is a particular pressure distribution along the tubing. This pressure-depth profile is called a pressure traverse and is shown in Figure 2.5. Gas liberation, gas expansion, and oil shrinkage along the production tubing can be treated as a series

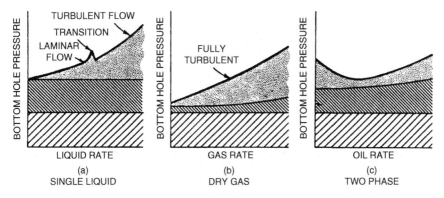

FIGURE 2.4   Components of pressure losses in tubing [2].

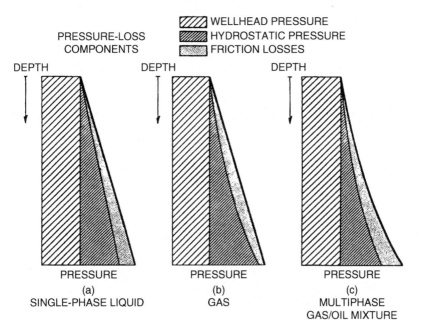

FIGURE 2.5   Pressure traverse for single-phase liquid, gas, and multiphase gas–oil mixture [2].

of successive incremental states in which saturated oil and gas coexist in equilibrium (flash process). This model is shown in Figure 2.6. At (a) the single-phase oil enters the wellbore; (b) marks the first evolution of gas, at the mixture's bubble point; and (c) and (d) show the traverse into the two-phase region. Note that the gas and oil P-T diagrams describing equilibrium phases at points c and d are not the same. This means the composition of equilibrium gas and oil phases changes continuously in the two-phase region. As the two-phase region is entered and gas is liberated, oil and gas phases change in volume and composition, but they are always in a saturated state, the gas at its dew point, and the oil at its bubble point. In Figure 2.7 the separation process is shown in forms of the resulting gas and oil.

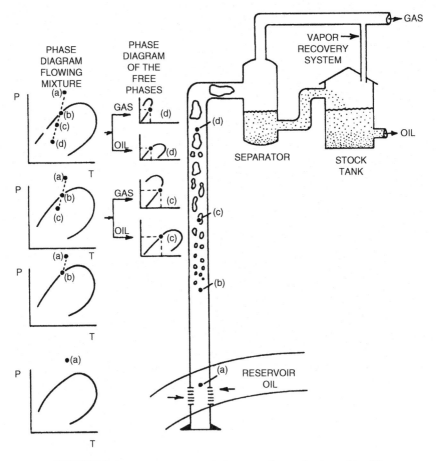

FIGURE 2.6   Changes in phase behavior in the production tubing [2].

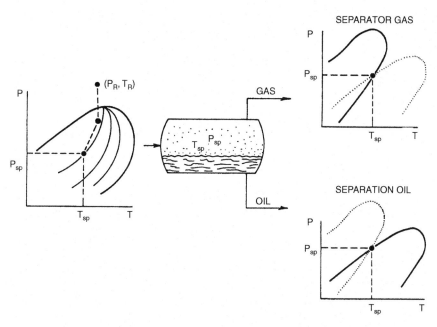

FIGURE 2.7 Pressure–temperature phase diagram used to describe surface separation [2].

Engineering analysis of two-phase fluid flow in pipes has focused primarily on the problem of predictive pressure drop, or pressure gradient, from Equation 2.6.

In many cases, it has become possible to treat two-phase pipeline problems with empirical numerical techniques that yield reasonably accurate pressure drops. Most of two-phase pipeline simulation currently is performed using "black oil" simulators. A black oil model's validity rests on the assumption that the hydrocarbon mixture is composed of two phases, denoted oil and gas, each with fixed composition. A black-oil model usually treats P–V–T properties (solution gas, densities, and viscosities) as single-value function of pressure. More sophisticated models include the temperature effect on fluid properties as well. The multicomponent or compositional approach is designed for gas condensate and volatile oil systems. These fluids are represented as N-component hydrocarbon mixtures, where N might be equal to components C_1, C_2, C_5, i-C_4, n-C_4, i-C_5, n-C_5, C_6, and C_{7+}. Equations of state (SRK, PR, SBWR) are used to determine physical properties. The term "compositional" implies that the overall or in situ fluid composition varies point by point with the distance as is shown in Figure 2.6. When a multicomponent gas–liquid mixture flows through a pipe, the composition, pressure, temperature and liquid holdup distributions are related [14].

2.1 BASIC PARAMETERS OF MULTIPHASE FLOW [1]

Knowledge of the flow regime determines the selection of the appropriate model for pressure gradient and liquid holdup. The flow regime,
pressure gradient, and liquid holdup are calculated for each segment of the
pipeline. The information needed to make the calculations includes:

1. pipeline inlet and outlet boundary conditions (liquid and gas flowrates,
   temperature, and pressure)
2. pipeline geometry, with segments specifications (any riser or well,
   down-comer, inclined section)
3. fluid properties (assume constant properties, compositional analysis,
   black oil approaches); this includes gas, oil, and water density; viscosity;
   and surface tension

It is assumed by flow regime that the distribution of each phase in the
pipe is relative to one another. Prediction of flow patterns for horizontal
flow is a more difficult task than for vertical flow. Possible flow regimes are
shown in Figure 2.8. An example of the complexity of two-phase flows in
Figure 2.8 shows a schematic sequence of flow patterns in vertical pipe
(Figure 2.9). Numerous authors [1–4] have presented flow-pattern and
flow-regime maps in which various areas are indicated on a graph for
which there are two independent coordinates. Maps are also dependent
on ranges of pipe inclination for vertical upward to vertical downward [1].
Selection of the appropriate flow regime map is based solely on the pipe
segment inclination θ from the horizontal, as shown below:

Range of Inclination from the Horizontal	Regime Maps
θ = 90° to 15°	Upward inclined
θ = 15° to −10°	Near horizontal
θ = −10° to −90°	Downward inclined

2.1.1 Flow Regimes

The steps in the determination of the flow regime are as follows:

1. Calculate dimensionless parameters.
2. Refer to flow regime maps laid out in coordinates of these parameters.
3. Determine the flow regime by locating the operating point on the flow
   regime map.

The discussions in the following sections treat the flow regime maps for
vertical upward (θ = 90°); slightly inclined (θ = 15° to −10°) and vertical
downward inclinations.

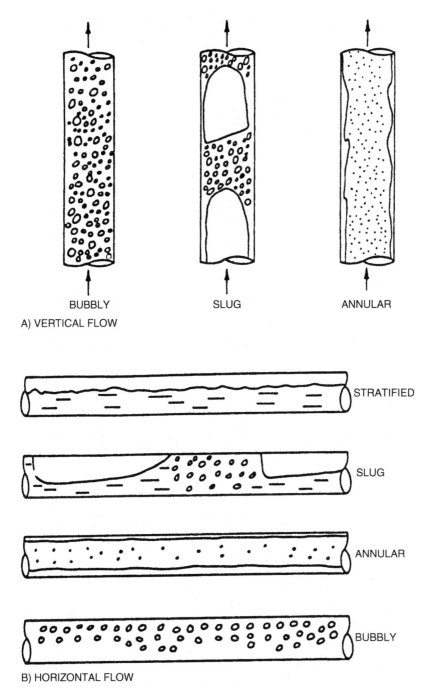

A) VERTICAL FLOW

B) HORIZONTAL FLOW

FIGURE 2.8   Gas–liquid flow regimes.

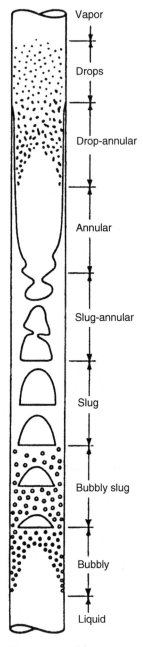

FIGURE 2.9   Possible sequence of flow patterns in a vertical tube [3].

To proceed with the calculations of the flow regime, it is necessary to calculate the superficial velocities for each flow phase. The superficial velocities for the gas, oil and water are

$$V_{sg} = q_g/A_p \tag{2.7a}$$

$$V_{so} = q_o/A_p \tag{2.7b}$$

$$V_{sw} = q_w/A_p \tag{2.7c}$$

where A_p = pipe flow area in ft^2
q = volumetric flowrate at flow conditions in ft^3/s

The superficial velocities of liquid phase (oil and water) are calculated as

$$V_{sL} = V_{so} + V_{sw} \tag{2.7d}$$

The mixture velocity that will be used in some of the calculations is the sum of the superficial velocities of the gas and the liquid phases:

$$V_m = V_{sL} + V_{sg} \tag{2.7e}$$

The average velocity of each phase is related to the superficial velocity through the liquid holdup:

$$U_L = v_L = v_{sL}/H_L \tag{2.7f}$$

$$U_g = v_g = v_{sg}/(1 - H_L) \tag{2.7g}$$

For a homogeneous model, both phases are assumed to have equal velocities and each is equal to a two-phase (or mixture) velocity:

$$v_L = v_g = v_m \tag{2.7h}$$

H_L in Equations 2.7f and 2.7g refers to liquid holdup.

Liquid holdup is defined as the ratio of the volume of a pipe segment occupied by liquid to the volume of the pipe segment:

$$H_L = \frac{\text{volume of liquid in a pipe segment}}{\text{volume of pipe segment}} \tag{2.8a}$$

In some cases, e.g., for stratified horizontal flow regime, liquid holdup can be calculated as follows:

$$H_L = \frac{A_L}{A_L + A_g} \tag{2.8b}$$

where A_L = cross-sectional area occupied by liquid (oil and water)
A_g = cross-sectional area occupied by gas

2.2 SLIGHTLY INCLINED PIPES $(-10° < \theta < 15°)$

As can be seen from Figure 2.8, there are four flow regimes of interest—stratified, slug, annular, and bubbly—and three flow regime transition zones.

2.2.1 Step 1. Dimensionless Parameters

1. Martinelli parameter—this is the ratio of the liquid and gas phases as if each flowed alone in the pipe:

$$X = \left[\frac{(dp/dL)_{Ls}}{(dp/dL)_{gs}} \right]^{0.5} = \left(\frac{2f_{hLs}\gamma_L v_{sL}^2/D}{2f_{wgs}\gamma_s v_{sg}^2/D} \right) \tag{2.9a}$$

According to standard fluid mechanics book

$$f(*) = \begin{cases} 0.046/Re^{0.2} & \text{if } Re = \gamma Dv/\mu > 1500 \\ 16/Re & \text{if } Re < 1500 \end{cases} \tag{2.9b}$$

2. Gas Froude number (dimensionless gas flowrate):

$$F_g = v_{sg} \left(\frac{\gamma_g}{(\gamma_L - \gamma_g)gD} \right)^{0.5} \tag{2.10}$$

3. Turbulence level

$$T = \left[\frac{(dp/dL)_{Ls}}{g(\gamma_L - \gamma_g)\cos\theta} \right] = \left[\frac{2f_{wLs}\gamma_L v_{sL}^2/D}{g(\gamma_L - \gamma_g)\cos\theta} \right] \tag{2.11}$$

4. Dimensionless inclination (slope parameter)

$$Y = \left[\frac{g(\gamma_L - \gamma_g)\sin\theta}{(dp/dL)_{gs}} \right] = \left[\frac{g(\gamma_L - \gamma_g)\sin\theta}{2f_{wgs}\gamma_g v_{sg}^2/D} \right] \tag{2.12}$$

2.2.2 Step 2. Flow Regime Map

The parameter Y is used to select the flow regime map to be used from Figure 2.10a–i. This figure presents nine flow regime maps (a to i) for a wide range of dimensionless inclinations, Y, in the range of interest.

---

*Calculated separately for "ls" and "gs."

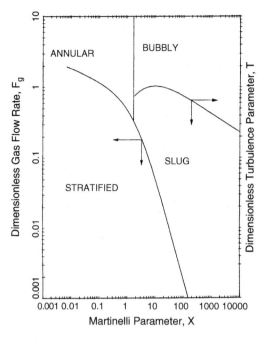

FIGURE 2.10a   Flow regime map for slightly inclined pipes (horizontal) [4].

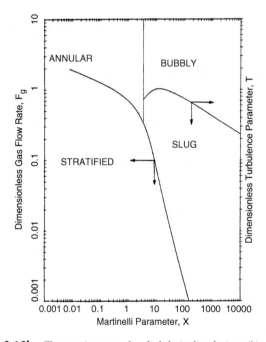

FIGURE 2.10b   Flow regime map for slightly inclined pipes $(Y = -100)$ [4].

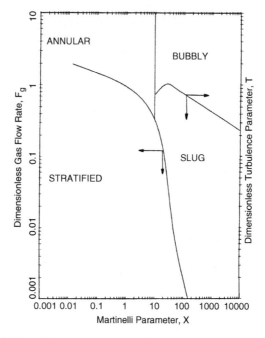

FIGURE 2.10c   Flow regime map for slightly inclined pipes $(Y = -1000)$ [4].

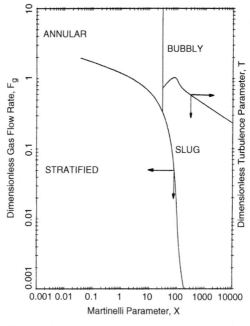

FIGURE 2.10d   Flow regime map for slightly inclined pipes $(Y = -10,000)$ [4].

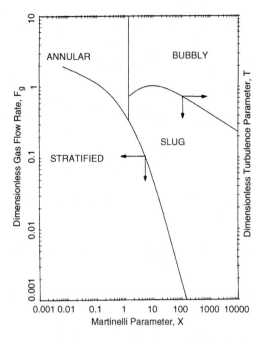

FIGURE 2.10e Flow regime map for slightly inclined pipes $(Y = 10)$ [4].

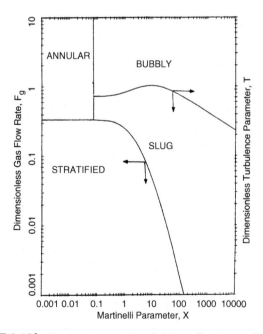

FIGURE 2.10f Flow regime map for slightly inclined pipes $(Y = 30)$ [4].

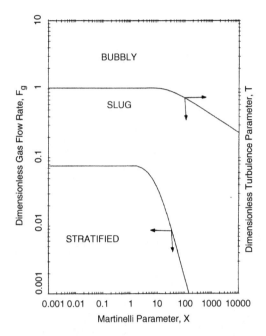

FIGURE 2.10g   Flow regime map for slightly inclined pipes (Y = 100) [4].

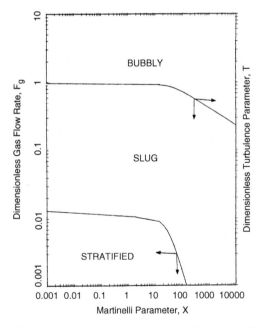

FIGURE 2.10h   Flow regime map for slightly inclined pipes (Y = 1000) [4].

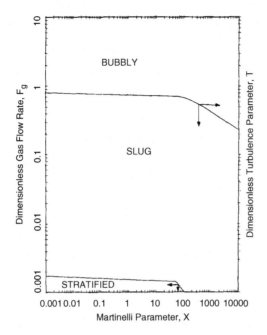

FIGURE 2.10i   Flow regime map for slightly inclined pipes (Y = 10,000) [4].

2.2.3 Step 3. Flow Regime Selection

The flow regime maps are prepared in Froude number–Martinelli parameter–turbulence level (F_g–X–T). Four flow regimes are noted on the maps, doing with the three transition boundaries. Regimes are stratified, slug, annular, and bubbly. The flow conditions for the current pipe segment are located using the X, F_g coordinates. If the located point is in the region labeled "stratified," then the flow regime is indeed stratified. If the located point is outside of the stratified region, then determine whether the point is to the left or right of the vertical line representing the transition between annular and slug flow regimes. If the point X, F_g is on the right side of the vertical line, X,T coordinates are necessary.

2.3 RISERS AND WELLS (θ = 90°)

There are three possible flow regimes, including annular, slug, and bubbly, and two regime transitions.

2.3.1 Step 1. Dimensionless Parameters

For this regime map, three dimensionless parameters are calculated. The parameters include dimensionless groups, which represent a balance between buoyancy, inertial, and surface tension forces. There are called Kutateladze numbers.

$$K_g = \frac{\gamma_g^{0.5} v_{sg}}{\left[g\sigma_{go}(\gamma_L - \gamma_g)\right]^{0.25}} \tag{2.13}$$

$$K_L = \frac{\gamma_L^{0.5} v_{sL}}{\left[g\sigma_{go}(\gamma_L - \gamma_g)\right]^{0.25}} \tag{2.14}$$

$$\rho^* = \left(\frac{\rho_L}{\rho_g}\right) \tag{2.15}$$

where K_g, K_L = dimensionless gas and liquid flowrates
$\quad\rho_g, \rho_L$ = density of gas and liquid phases in $lb_m/ft.^3$
$\quad v_{sg}, v_{sL}$ = superficial gas and liquid velocities in ft./s
$\quad\sigma_{go}$ = surface tension between gas–oil phases in $lb_f/ft.$
$\quad\rho^*$ = dimensionless density ratio
$\quad g$ = acceleration of gravity in ft./s^2

2.3.2 Step 2. Flow Regime Map

Once these dimensionless parameters are calculated, then the point with the coordinates (K_g, K_L) is located on the flow regime map in Figure 2.11.

2.3.3 Step 3. Flow Regime Selection

The boundary for the appropriate density ratio is located for the transition boundary between the slug and bubbly flow regimes. Once the appropriate boundary line is found, then the flow regime is simply bubbly, slug, or annular depending upon the region in which the point falls. This slug to annular transition applies only if the pipe size D is larger than a critical diameter D_{crit} given by

$$D_{crit} = 1.9 \left[\frac{\sigma_{go}(\gamma_L - \gamma_g)}{g\gamma_L}\right]^{0.25} N_L^{(-0.4)} \tag{2.16a}$$

where

$$N_L = \frac{\mu_L}{(\gamma_L\sigma_{go})^{0.5} \left[\sigma_{go}/g(\gamma_L - \gamma_g)\right]^{0.25}} \tag{2.16b}$$

μ_L = liquid phase viscosity in $lb_m/(ft./s)$

Usually, the critical pipe size is about 2 in. for conditions of gas and oil pipelines so that Figure 2.11 can be used often. The criterion should be checked each time, however. If $D < D_{crit}$, another method has to be used; see Reference [22], Vol. 3.

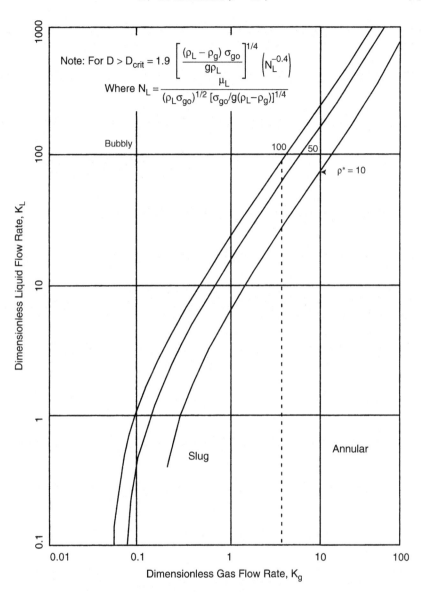

Note: For $D > D_{crit} = 1.9 \left[\dfrac{(\rho_L - \rho_g)\, \sigma_{go}}{g\rho_L} \right]^{1/4} \left(N_L^{-0.4} \right)$

Where $N_L = \dfrac{\mu_L}{(\rho_L \sigma_{go})^{1/2} \, [\sigma_{go}/g(\rho_L - \rho_g)]^{1/4}}$

FIGURE 2.11   Flow regime map for vertical upward inclinations ($\theta = 90°$) [4].

2.4 DOWNCOMERS ($\theta = -90°$)

There are also three possible flow regimes: annular, slug, and bubbly. There are two flow regime transitions to be calculated. Two different maps will be used: one for transition between annular and slug flow regimes, and a second for the transition between the slug and bubbly flow regimes.

2.4.1 Annular-Slug Transition

1. Dimensionless parameters X and |Y| from Equations 2.9a and 2.12, respectively.
2. Flow regime map, see Figure 2.12.
3. Flow regime selection. Locate the point with (X, |Y|) coordinates; if the point falls in the region "annular," then the flow regime is annular. If the point falls in region "slug" or "bubbly," then the map from Figure 2.13 must be used.

2.4.2 Slug-Bubbly Transition

1. Calculate dimensionless parameters K_g, K_L, and ρ^*; use Equations 2.14 and 2.15, respectively.
2. The appropriate regime boundary for the specific weight ratio γ^* is selected on the regime map with K_g, K_L coordinates.
3. Flow regime selection. Depending upon which region of the map the data point falls into, flow regime is slug or bubbly.

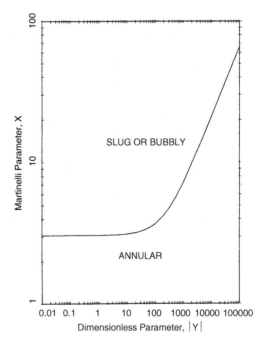

FIGURE 2.12   Flow regime map for slug-annular transition for vertical downward inclination ($\theta = -90°$) [4].

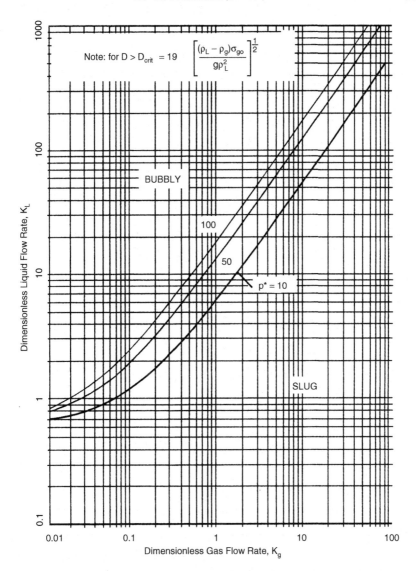

FIGURE 2.13 Flow regime map for bubble-slug transition for vertical downward inclinations ($\theta = -90°$) [4].

Similarly as for ($\theta = 90°$), Figure 2.13 applies for pipes greater than a certain critical diameter.

$$D > D_{crit} = 19 \left[\frac{(\gamma_L - \gamma_g)\sigma_{go}}{g\gamma_L^2} \right]^{0.5} \tag{2.17}$$

The above procedure is valid for $D > 2$ in.

For each flow regime, there is a separate pressure gradient and holdup calculation method.

2.5 STRATIFIED FLOW REGIME

First, the *liquid holdup* H_L and f_i/f_{wg} (friction factor ratio) are calculated. The liquid holdup as a function of X (Equation 2.9a), Y (Equation 2.12), and (f_i/f_{wg}) is read from Figure 2.14a–d. $f_i/f_{wg} = 10$ is recommended to be used as a preliminary estimate. For better accuracy f_i/f_{wg} can be calculated. H_L

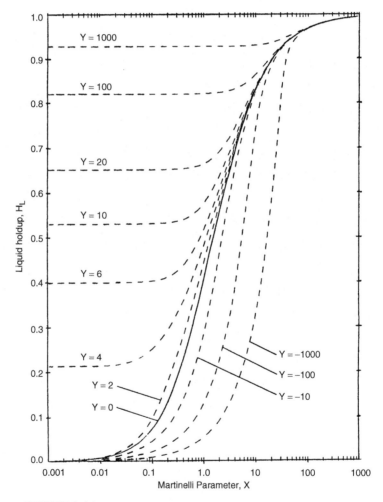

FIGURE 2.14a Liquid holdup in stratified flow regime ($f_i/f_{wg} = 1$) [4]

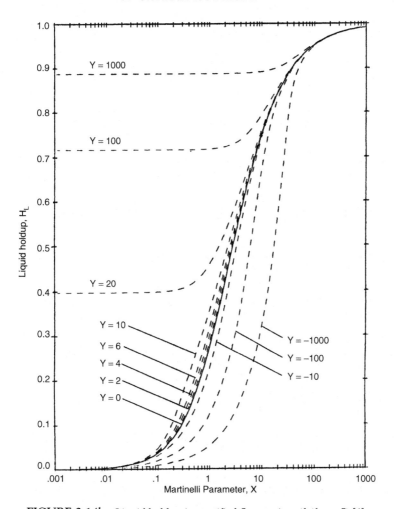

FIGURE 2.14b   Liquid holdup in stratified flow regime ($f_i/f_{wg} = 5$) [4].

should be first estimated by the method described above and

$$\frac{f_i}{f_{wg}} = 1.0 \text{ if Equation 2.18b} < 1.0 \tag{2.18a}$$

$$\frac{f_i}{f_{wg}} = \left[2 + \frac{0.000025\text{Re}_{Ls}}{D/3.281}\right](1 - H_L)^{5/2}(1 + 75H_L) \tag{2.18b}$$

if Equation 2.18b > $(1 + 75H_L)$ \hfill (2.18c)

(D is in inches).

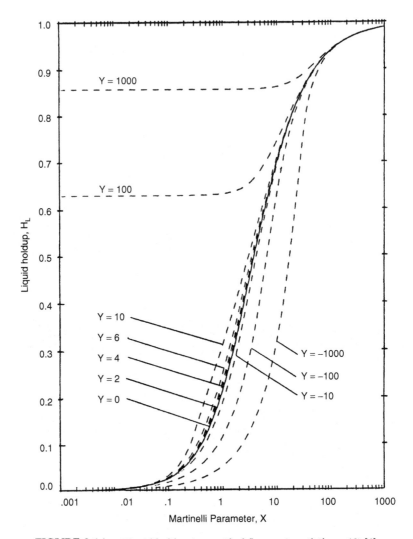

FIGURE 2.14c   Liquid holdup in stratified flow regime ($f_i/f_{wg} = 10$) [4].

Figure 2.15a–h shows Equation 2.18a for pipe diameters ranging from 4 to 36 in. The friction factor ratio (f_i/f_{wg}) can be estimated from the plots in Figure 2.15a–h or as a result of calculation.

When the friction factor is determined correctly, Figure 2.14 should again be used to estimate a new value of the liquid holdup. If the new value of the liquid holdup obtained is different from determined previously, this new value should be used in Equation 2.18a or Figure 2.15a–h to refine the estimate for f_i/f_{wg}. This is simply an iteration process.

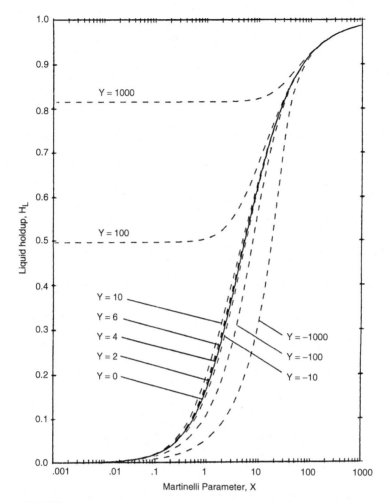

FIGURE 2.14d Liquid holdup in stratified flow regime ($f_i/f_{wg} = 20$) [4].

2.5.1 Pressure Gradient

Knowing the liquid holdup, H_L, calculations for the pressure gradients due to friction and gravitational effects are straightforward. First, some geometric parameters are calculated:

1. Dimensionless cross-sectional area occupied by gas (A_G) and by liquid (A_L)

$$A_G = 0.25[\cos^{-1}(2h^* - 1) - (2h^* - 1)\left[1 - (2h^* - 1)^2\right]^{0.5} \qquad (2.19a)$$

$$A_L = 0.25[\pi - \cos^{-1}(2h^* - 1) + (2h^* - 1)\left[1 - (2h^* - 1)^2\right]^{0.5} \qquad (2.19b)$$

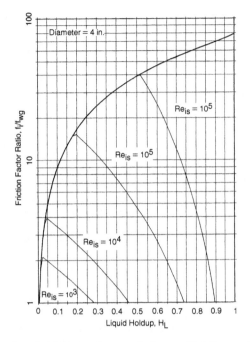

FIGURE 2.15a   Interfacial friction factor ratio for stratified flow regime (D = 4 in.) [4].

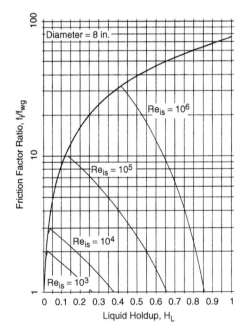

FIGURE 2.15b   Interfacial friction factor ratio for stratified flow regime (D = 8 in.) [4].

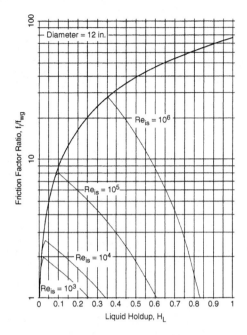

FIGURE 2.15c Interfacial friction factor ratio for stratified flow regime (D = 12 in.) [4].

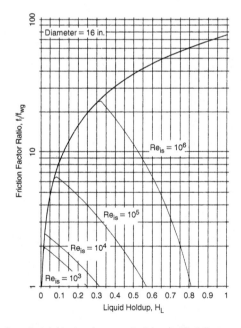

FIGURE 2.15d Interfacial friction factor ratio for stratified flow regime (D = 16 in.) [4].

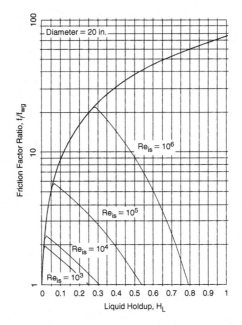

FIGURE 2.15e   Interfacial friction factor ratio for stratified flow regime (D = 20 in.) [4].

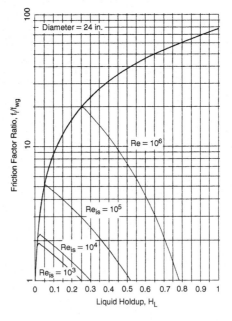

FIGURE 2.15f   Interfacial friction factor ratio for stratified flow regime (D = 24 in.) [4].

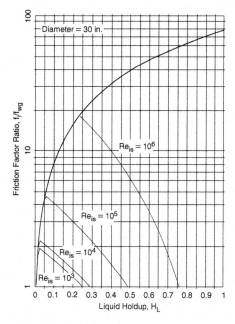

FIGURE 2.15g Interfacial friction factor ratio for stratified flow regime (D = 30 in.) [4].

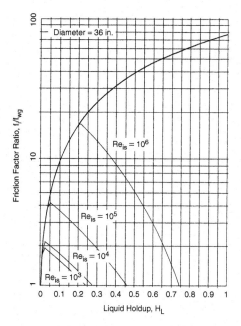

FIGURE 2.15h Interfacial friction factor ratio for stratified flow regime (D = 36 in.) [4].

2. Dimensionless wetted perimeter for gas phase (S_G) and for liquid phase (S_L)

$$S_G = \cos^{-1}(2h^* - 1) \tag{2.20a}$$

$$S_L = \left[\pi - \cos^{-1}(2h^* - 1) \right] \tag{2.20b}$$

3. Dimensionless interfacial length between gas and liquid phases (S_i)

$$S_i = \left[1 - (2h^* - 1) \right]^{0.5} \tag{2.21}$$

4. Dimensionless hydraulic diameter for gas phase (D_G) and liquid phase (D_L)

$$D_G = 4A_G/(S_L + S_i) \tag{2.22a}$$

$$D_L = 4A_L/S_L \tag{2.22b}$$

where h = dimensionless liquid in pipe ($= H_L/D$)

Dimensionless liquid level h can be expressed in terms of H_L as shown in Figures 2.16 to 2.19, but in terms of h that approach is much easier for the circular pipe cross-section.

Since the liquid holdup is

$$H_L = 4A_L/\pi, \tag{2.23}$$

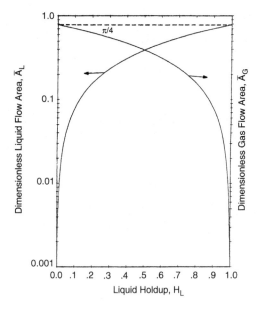

FIGURE 2.16 Dimensionless cross-sectional areas in stratified flow regimes [4].

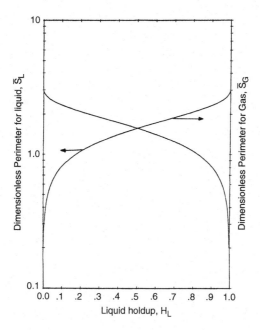

FIGURE 2.17 Dimensionless wetted perimeters in stratified flow regimes [4].

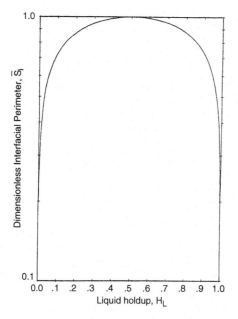

FIGURE 2.18 Dimensionless interfacial perimeters in stratified flow regimes [4].

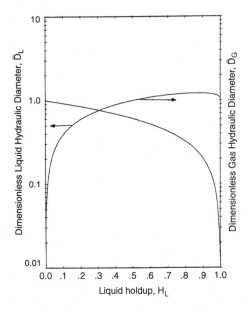

FIGURE 2.19   Dimensionless hydraulic diameters in stratified flow regimes [4].

It is necessary to find h first, using the expression in Equations 2.19 to 2.22. The value of h must be found numerically, or alternatively, it can be estimated from Figure 2.20. After geometric parameters are calculated (Equations 2.19 to 2.22), the friction factors for the gas and liquid phase has to be calculated:

$$f_{wg} = \begin{cases} 0.046/Re_g^{0.2} & \text{if } Re_g = \left(\frac{\gamma_g D_G D v_{sg}}{\mu_g(1-h_L)}\right) \geq 1500 \\ 16/Re_g & \text{if } Re_g < 1500 \end{cases} \tag{2.24a}$$

$$f_{wL} = \begin{cases} 0.046/Re_L^{0.2} & \text{if } Re_L = \left(\frac{\gamma_g D_L D v_{sL}}{\mu_L H_L}\right) \geq 1500 \\ 16/Re_L & \text{if } Re_L < 1500 \end{cases} \tag{2.24b}$$

Then the frictional and gravitational pressure gradients given by Expression 2.6 are

$$\left(\frac{dp}{dL}\right)_f = -\left[2f_{wL}\left(\frac{\gamma_L v_{sL}^2}{\pi D}\right)\left(\frac{1}{H_L}\right)^2 (S_L)\right]$$

$$-\left[2f_{wg}\left(\frac{\gamma_g v_{sg}^2}{\pi D}\right)\left(\frac{1}{1-H_L}\right)(S_G)\right] \tag{2.25a}$$

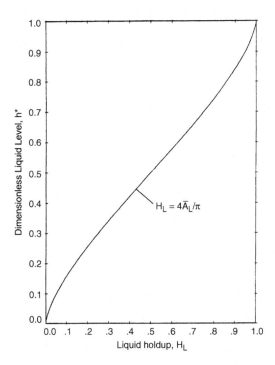

FIGURE 2.20 Dimensionless liquid level stratified flow regimes [4].

$$\left(\frac{dp}{dL}\right)_{el} = \left(\frac{dp}{dZ}\right)\sin\theta = -g\left[H_L\gamma_L + (1-H_L)\gamma_g\right]\sin\theta \qquad (2.25b)$$

where L = distance in ft.

z = vertical coordinate in ft.

S = pipeline inclination from the horizontal in degree

2.5.2 Special Cases for Low and High Liquid Holdup

As $H_L \approx 1.0$ or 0.0, it becomes difficult to determine the liquid holdup accurately. Small errors in the estimation lead to large errors in the frictional portion of the pressure gradient by Equation 2.25a; therefore, in such cases as above, it is recommended that the pressure gradient be calculated by the following methods. If the liquid holdup $H_L > 0.99$, then

$$\left(\frac{dp}{dL}\right)_t = -\left[f_{wL}\left(\frac{\gamma_L v_{sL}}{2D}\right)^2\left(\frac{1}{H_L}\right)^2\left(\frac{S_L}{A_L}\right)\right]$$

$$+ \left[\left(\frac{f_i}{f_{wg}}\right)f_{wg}\left(\frac{\gamma_g v_{sg}^2}{2D}\right)\left(\frac{1}{1-H_L}\right)^2\left(\frac{S_i}{A_L}\right)\right] - g\gamma_L\sin\theta \qquad (2.26a)$$

If the liquid holdup $H_L < 0.01$, then

$$\left(\frac{dp}{dL}\right)_t = -\left\{ f_{wg} \frac{\gamma_g v_{sg}^2}{2D} \left(\frac{1}{1-H_L}\right)^2 \left[\left(\frac{S_G}{A_G}\right) + \left(\frac{f_i}{f_{wg}}\right)\right.\right.$$
$$\left.\left.\times \left(\frac{S_i}{A_G}\right)\right]\right\} - g\gamma_g \sin\theta \qquad (2.26b)$$

where the gravitational pressure gradient is still the same as in Equation 2.25b and the frictional pressure gradient can be determined by subtraction from Equation 2.26a or 2.26b.

2.6 ANNULAR FLOW REGIME

In the annular flow regime, the extent of liquid entrainment must first be estimated, then the liquid holdup and the pressure gradient can be calculated.

2.6.1 Liquid Entrainment E_d

Calculation methods in this area have not been validated and may be poor. First, calculation v_e (critical gas velocity on set of entrainment) is made as follows:

$$v_e = 0.00025 \left(\frac{\gamma_L}{\gamma_g}\right)^{0.5} \left(\frac{\sigma_{go}}{\mu_g}\right) \qquad (2.27)$$

If the value of the $v_{sg} < v_e$, then there is no entrainment $E_d = 0$. If $v_{sg} > v_e$, then the entrainment fraction should be estimated by

$$E_d = 1 - \exp\left[0.23\left(\frac{v_e - v_{sg}}{v_e}\right)\right] \qquad (2.28)$$

where E_d = mass fraction or volume fraction of the total liquid flow that is in the form of entrained droplets

Equation 2.28 is an empirical correlation without experimental basis. Figure 2.21 graphically shows the liquid entrainment fraction for various values of the critical entrainment velocity, v_e.

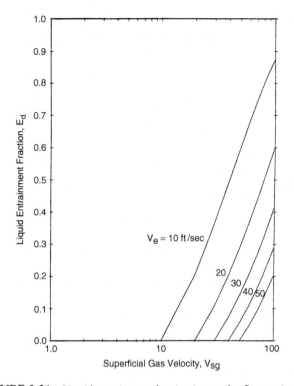

FIGURE 2.21 Liquid entrainment fraction in annular flow regime [4].

2.6.2 Liquid Holdup H_L

After E_d fraction is calculated, the liquid holdup can be estimated. To determine H_L dimensionless specific weight ratio γ_c^* has to be known:

$$\gamma_c^* = \frac{\gamma_c}{\gamma_g} = \left[E_d \left(\frac{\gamma_L}{\gamma_g} \right) \left(\frac{v_{sL}}{v_{sg}} \right) + 1 \right] \tag{2.29}$$

$\gamma_c^* = (\text{density of a gas} - \text{droplet mixture in the core of the annular flow}) / (\text{density of the gas phase})$

with this density ratio, other dimensionless parameters could be defined:

$$X_a = \frac{X}{(\gamma_c^*)^{0.5}} \tag{2.30a}$$

$$Y_a = \left(\frac{Y}{\gamma_c^*} \right) \left(\frac{\gamma^* - \gamma_c^*}{\gamma^* - 1} \right) \tag{2.30b}$$

where X and Y are the same as the values for the stratified flow regime.

With these parameters, Figure 2.22a–d can be used to estimate the liquid holdup in the liquid film H_{Lf} for the annular flow regime.

The total liquid holdup in the annular flow regime is calculated as follows:

$$H_L = H_{Lf} + \left(\frac{E_d v_{sL}}{E_d v_{sL} + v_{sg}} \right) \tag{2.31}$$

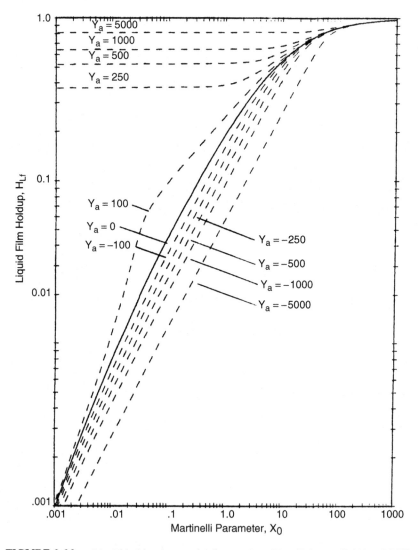

FIGURE 2.22a Liquid holdup in annular flow regime ($E_d = 0$) (expanded Y scale) [4].

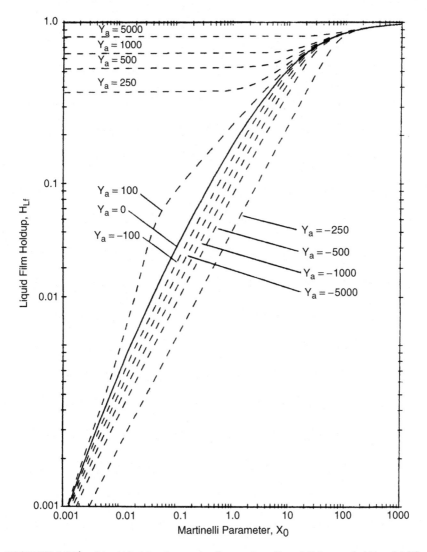

FIGURE 2.22b Liquid holdup in annular flow regime ($E_d = 0.2$) (expanded Y scale) [4].

2.6.3 Pressure Gradient

The friction factor for the liquid phase needs to be calculated first.

$$f_{wL} = \begin{cases} 0.046/Re_L^{0.2} & \text{if } Re_L = \left[\dfrac{\gamma_g D v_{sL}(1 - E_d)}{\mu_L} \right] \geq 1500 \\ 16/Re_L & \text{if } Re_L < 1500 \end{cases} \qquad (2.32)$$

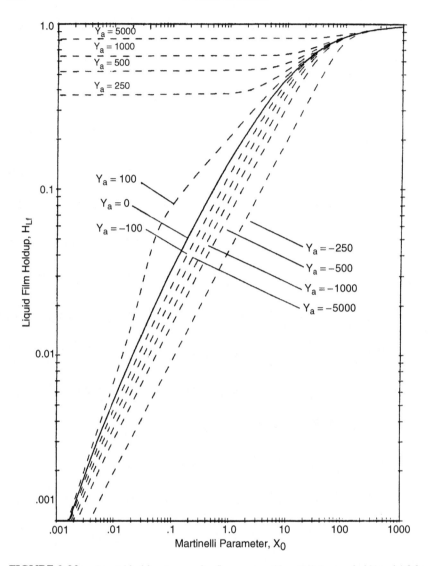

FIGURE 2.22c   Liquid holdup in annular flow regime ($E_d = 0.4$) (expanded Y scale) [4].

The fractional and gravitational pressure gradients are

$$\left(\frac{dp}{dL}\right)_f = -\left[2f_{wL}\left(\frac{\gamma_L v_{sL}^2}{D}\right)\left(\frac{1-E_d}{H_{Lf}}\right)^2\right] \tag{2.33a}$$

and

$$\left(\frac{dp}{dL}\right)_g = g\left[\gamma_L H_{Lf} + (1-H_{Lf})\gamma_c\right]\sin\theta \tag{2.33b}$$

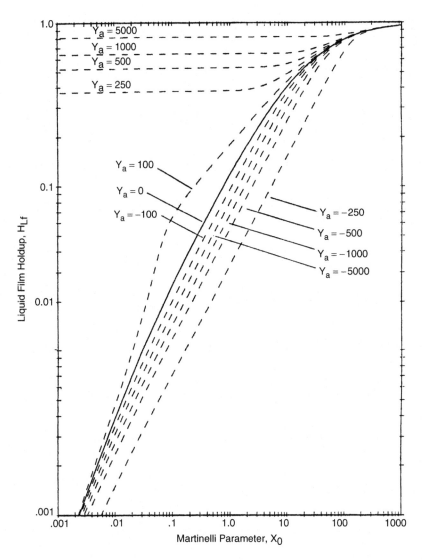

FIGURE 2.22d   Liquid holdup in annular flow regime ($E_d = 0.6$) (expanded Y scale) [4].

2.6.4 Special Case for Low Liquid Holdup

For low values of the liquid holdup, the approach described below is recommended. If $H_{Lf} < 0.002$, then

$$f_{wg} = \begin{cases} 0.046/Re_g^{0.2} & \text{if } Re_g = \left[\dfrac{\gamma_g D v_{sg}}{\mu_g(1-H_{Lf})}\right] \geq 1500 \\[2mm] 16/Re_g & \text{if } Re_g < 1500 \end{cases} \qquad (2.34a)$$

Then the total pressure gradient is

$$\left(\frac{dp}{dL}\right)_{total} = -2(1+75H_{Lf})(f_{wg})\left(\frac{\gamma_c v_{sg}^2}{D}\right)\left(\frac{1}{1-H_{Lf}}\right)^{5/2}$$
$$- g\gamma_c \sin\theta \qquad (2.34b)$$

$(dp/dL)_g$ from Equation 2.33b could be calculated.

2.7 SLUG FLOW REGIME

In slug flow, the liquid slugs tend to contain some gas. The fraction of liquid in the liquid slugs can be estimated by

$$E_{Ls} = \frac{1}{1+\left(\dfrac{v_m}{28.4}\right)^{1.39}} \qquad (2.35)$$

Figure 2.23 shows the liquid holdup versus the mixture velocity v_m by this correlation.

2.7.1 Slug Velocity

The velocity of the liquid slugs are gas bubbles is determined by

$$v_s = C_o v_m + k\left[\frac{gD(\gamma_L - \gamma_g)}{\gamma_L}\right]^{0.5} \qquad (2.36)$$

For C_o and k, see Table 2.1.

2.7.2 Liquid Holdup

The overall liquid holdup is obtained by

$$H_L = 1 - \frac{v_{sg} + (1 - E_{Ls})(v_s - v_m)}{v_s} \qquad (2.37)$$

TABLE 2.1   Drift-Flux Parameters for Slug Flow Regime

Pipe Inclination	C_o	k
$80° < \theta \leq 90°$	1.2	0.35
$0° < \theta \leq 80°$	1.3	0.50
$\theta = 0 \leq 1 < 1$	1.3	0
$-80° < \theta < 0°$	1.3	−0.5*
$-90° < \theta < -80°$	0.9	−0.6*

*If the slug velocity becomes small as a result of having the mixture superficial velocity, v_m, very small, then the slug velocity, v_s, should be limited to the mixture velocity.

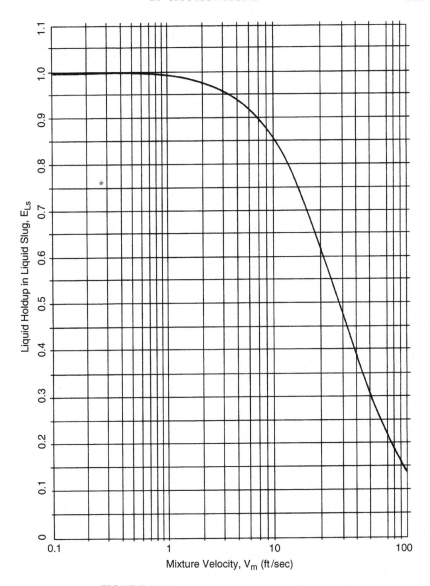

FIGURE 2.23   Liquid holdup in liquid slug [4].

2.7.3 Pressure Gradient

Two cases will be considered, slightly inclined flow and vertical flow. Pressure for slightly inclined flow is a function of an average liquid velocity v_L, a friction factor f_L, and an average slug density ρ_{Ls}.

$$v_L = v_{sL}/H_L \qquad (2.38a)$$

$$f_L = \begin{cases} 0.046/Re_L^{0.2} & \text{if } Re_L = \left(\dfrac{\gamma_L D v_L}{\mu_L} \right) \geq 1500 \\[2mm] 16/Re_L & \text{if } Re_L < 1500 \end{cases} \qquad (2.38b)$$

$$\gamma_{sL} = \left[E_{Ls}\gamma_L + (1 - E_{Ls})\gamma_g \right] \qquad (2.38c)$$

The frictional pressure gradient is then calculated by

$$\left(\frac{dp}{dL} \right)_f = - \left(\frac{2f_L \gamma_{Ls} v_L^2}{D} \right) \qquad (2.39a)$$

and the gravitational pressure gradient is

$$\left(\frac{dp}{dL} \right)_g = -g\left[H_L\gamma_L + (1 - H_L)\gamma_g \right]\sin\theta \qquad (2.39b)$$

For vertical flow, the fractional pressure gradient is calculated by

$$\left(\frac{dp}{dL} \right)_f = - \left[2f_m \left(\frac{\gamma_{Ls} v_m^2}{D} \right) H_L \right] \qquad (2.40)$$

where γ_{Ls}, from Equation 2.38c, with $E_{Ls} = 0.75$

$$H_L = \frac{v_{sL}}{v_L} = \frac{\text{liquid superficial velocity}}{\text{avg. liquid velocity in slug flow}}$$

and f_m, from Equation 2.38b, using the mixture velocity v_m (replacing v_L) in the Reynolds number.

2.7.4 Optional Correction

The approximation to the pressure gradient above neglects the liquid holdup in the liquid film around the gas bubble. This holdup may be significant for long gas bubbles that occur in wells of gas and oil pipelines. Thus Equations 2.40 and 2.39b will overpredict the pressure gradient. If greater accuracy is desired, a closer estimate (which may tend to underpredict the pressure gradient) is possible.

The thickness δ^* (dimensionless thickness of liquid film in slug bubble) of the liquid film is estimated and the result used to modify the liquid holdup. To get the liquid film thickness, two dimensionless parameters are first calculated; dimensionless velocity ratio for slug flow in risers N_f and dimensionless velocity ratio for slug flow in risers v^*:

$$N_f = \left[D^3 g(\gamma_1 - \gamma_g)\gamma_L \right]^{0.5} / \mu_L \qquad (2.41a)$$

$$v^* = v_m \gamma_L^{0.5} / k\left[gD(\gamma_L - \gamma_g) \right]^{0.5} \qquad (2.41b)$$

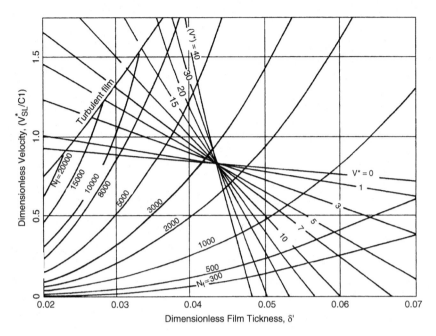

FIGURE 2.24   Dimensionless film thickness for well in slug flow [4].

In Figure 2.24, the line of constant v^* is found first; then the line of constant N_f is located. The "turbulent film" line is a limiting case for large values of N_f. The intersection of these two lines determines the film thickness δ^*. The liquid holdup calculated by Equation 2.37 is then modified by

$$H'_L = 1 - (1 - H_L)/(1 - 2\delta^*)^2 \tag{2.42}$$

This modified value of the liquid holdup is used in Equation 2.40 and 2.39b to determine the pressure gradient, where the other parameters are calculated as before.

2.8 BUBBY FLOW REGIME

The *liquid holdup* is found by a drift-flux model

$$H_L = 1 - \left(\frac{v_{sg}}{C_o v_m + k v_\infty}\right) \tag{2.43a}$$

where the velocity v_∞ is

$$v_\infty = \left[\frac{g \sigma_{go}(\gamma_L - \gamma_g)}{\gamma_L^2}\right]^{0.25} \tag{2.43b}$$

and the parameters C_o and k are determined as below:

Pipe Inclination	C_o	k
$> 0°$	$1.2 - 0.2(\gamma_g/\gamma_L)^{0.5}$	1.4
$0°$	$1.2 - 0.2(\gamma_g/\gamma_L)^{0.5}$	0
$< 0°$	0.9	0

2.8.1 Pressure Gradient

The friction factor for the liquid is first calculated by

$$f_{wLm} = \begin{cases} 0.046/Re_{Lm}^{0.2} & \text{if } Re_{Lm} = \left(\frac{\gamma_L D v_m}{\mu_g}\right) \geq 1500 \\ 16/Re_{Lm} & \text{if } Re_{Lm} < 1500 \end{cases} \tag{2.44a}$$

The pressure gradient due to friction in the bubbly flow regime is evaluated from

$$\left(\frac{dp}{dL}\right)_f = \frac{2f_{wLm} v_m (\gamma_g v_{sg} + \gamma_L v_{sL})}{D} \tag{2.44b}$$

and the gravitational pressure gradient

$$\left(\frac{dp}{dL}\right)_g = -g\left[H_L \gamma_L + (1 - H_L)\gamma_g\right]\sin\theta \tag{2.44c}$$

2.9 CORRECTION FOR ACCELERATION EFFECTS

The methods used in this section so far neglect the contribution of acceleration effects to the pressure gradient.

The acceleration pressure gradient in pipeline flow is due to the changes in the fluid properties through the pipe segment. These changes include expansion of the gas phase, expansion of the liquid phase, and the changes in quality (due to phase behavior).

To account for acceleration effects, the total pressure gradient should be calculated by:

$$\left(\frac{dp}{dL}\right)_t = \frac{\left(\frac{dp}{dL}\right)_f + \left(\frac{dp}{dL}\right)_g}{1 + A} \tag{2.45a}$$

where the parameter A is calculated by the method appropriate for each flow regime as follows:

$$A + \begin{cases} = 0 \quad \text{to neglect acceleration effects} \\[2mm] = (\gamma_g V_{sg} + \gamma_L V_{sL})^2 \left[X \left(\dfrac{dV_g}{dp} \right) + (1-X) \left(\dfrac{dV_L}{dp} \right) \right] \\ \quad \text{for bubbly or slug flow regime} \\[2mm] = (\gamma_g V_{sg} + \gamma_L V_{sL})^2 \left[\dfrac{x^2}{(1-H_L)} \left(\dfrac{dV_g}{dp} \right) + \dfrac{(1-x)^2}{H_L} \left(\dfrac{dV_L}{dp} \right) \right] \\ \quad \text{for stratified or annular flow regime} \end{cases}$$ (2.45b)

d_g/d_p and d_L/d_p represent the change in specific volume for each phase with a change in pressure as evaluated from the fluid properties. The acceleration portion of the pressure gradient is calculated by

$$(dp/dL)_a = -A(dp/dL)_t$$ (2.45c)

2.10 LIMITATION

The described-above methods can be applied under the following conditions:

1. flow mass in pipe in constant (steady-state flow)
2. only the gas–liquid phase flow is considered; the liquid phase is treated as an oil
3. the temperature is constant and equal to the average temperature

The above procedure presents the multiphase methods in a simplified form to permit quick prediction to be made. The same problems can be solved based on computer methods that are presented in Volumes 2 and 3 of the AGA project [1].

Example 2.1
Calculate the pressure gradient for a horizontal pipeline using the flow regime maps. The following data are given:

$q_g = 40\,\text{MMscf/d}$
$q_0 = 40{,}000\,\text{stb/d}$
$ID = 9\,\text{in.}$
$API(\text{gravity}) = 33°$
$P_{avg} = 2000\,\text{psia}$
$T_{avg} = 80°F$
$S.G. = 0.75$ at $14.7\,\text{psia}$ and $T = 60°F$
$R_p = 990\,\text{scf/bbl}$

Also calculate missing fluid properties using proper correlations.

$dp/dL = ?$

1. Stock-tank oil specific weight γ_{sto}

$$\gamma_{sto} = 62.37[141.5/(131.5 + °API)]$$
$$= 62.37[141.5/(131.5 + 33)]$$
$$= 53.65\,lb/ft.^3$$

2. Gas specific weight at std conditions γ_{gsc}

$$\gamma_{gsc} = 0.0763\gamma_g = 0.0763(0.75) = 0.05723\,lb/ft.^3$$

3. Pipe flow area

$$A_p = (\pi/4)D^2 = \pi/4(9/12)^2 = 0.4418\,ft.^2$$

4. Calculate Z, μ_g, μ_0, and GOR at P_{avg} and T_{avg}. From Gas P–V–T Program at $p = 2000\,psia$

$$T = 80°F$$
$$Z = 0.685$$
$$\mu_g = 0.0185\,cp$$

From Oil P–V–T program at $p = 2000\,psia$

$$T = 80°F$$
$$R_p = 990\,scf/bbl$$
$$\mu_0 = 2.96\,cp$$

5. Calculate γ_g at p_{avg}, T_{avg}, and Z_{avg}

$$\gamma_g = \frac{28.97(\gamma)p_{avg}}{Z_{ave}RT_{avg}} = \frac{28.97(0.75)(2000)}{10.685(10.73)(80 + 460)} = 10.95\,lb/ft.^3$$

6. Calculate vapor superficial velocity, v_{sg} from Equation 2.7

$$v_{sg} = \frac{q_g\rho_{gsc}}{A_p\rho_g(24)3600} = \frac{(40 \times 10^6)0.05724}{0.4418 \times 10.95 \times 24 \times 3600} = 5.48\,ft./s$$

7. Calculate liquid superficial velocity v_{sL}

$$v_{sL} = \frac{q_L}{A_p} = \frac{40,000(42)}{(7.481)24(3600)0.4418} = 5.88$$

8. Reynolds number and friction factor based on the superficial velocity, from Equation 2.9b

a. Liquid

$$R_{eL} = \frac{\gamma_L D v_{sL}}{\mu_{Lg}} = \frac{(53.56)(9/12)(5.88)}{(2.86)2.0886 \times 10^{-5}(32.2)} = 118{,}851.7 > 1500$$

$$f_{wLs} = \frac{0.046}{R_{eL}^{0.2}} = \frac{0.046}{(118{,}851.7)^{0.2}} = 4.44 \times 10^{-3}$$

where 2.0886×10^{-5} is a unit conversion factor

b. Gas

$$R_{eg} = \frac{\gamma_g D v_{sg}}{\mu_{gg}} = \frac{(10.95)(9/12)(5.48)}{(0.0187)(2.0886 \times 10^{-5})(32.2)} = 3{,}623{,}080 > 1500$$

$$f_{wgs} = \frac{0.046}{R_{eg}^{0.2}} = \frac{0.046}{(3{,}623{,}080)^{0.2}} = 2.24 \times 10^{-3}$$

9. Martinelli parameter X from Equation 2.9a

$$X = \left(\frac{2f_{wLs}\gamma_L v_{sL}^2/D}{2f_{wgs}\gamma_g v_{sg}^2/D} \right)^{0.5} = \left[\frac{4.44 \times 10^{-3}(53.65)(5.88)^2}{2.24 \times 10^{-3}(10.95)(5.48)^2} \right]^{0.5} = 3.3$$

10. Dimensionless inclination Y from Equation 2.12

$$Y = \frac{g(\gamma_L - \gamma_g)\sin\theta}{2f_{wgs}\gamma_g v_g^2/D}$$

since $\sin\theta = 0$ (horizontal pipe)

$$Y = 0$$

11. Calculate gas Froude number

$$F_g = v_{sg} \left[\frac{\gamma_g}{(\gamma_L - \gamma_g)gD} \right]^{0.5} = 5.48 \left[\frac{10.95}{(53.65 - 10.95)32.2(9/12)} \right]^{0.5} = 0.56$$

12. Select flow regime for $X = 3.3$ and $Y = 0$. Figure 2.10a indicates slug flow.
13. Liquid holdup in the liquid slug, E_{Ls}, from Equation 2.7e

$$v_m = v_{sL} + v_{sg} = 5.88 + 5.48 = 11.36\,\text{ft./s}$$

from Equation 2.35

$$E_{sL} = \frac{1}{1 + \left(\dfrac{v_m}{28.4} \right)^{1.39}} = \frac{1}{1 + \left(\dfrac{11.36}{28.4} \right)^{1.39}} = 0.781$$

14. Slug velocity v_s from Equation 2.36

$$v_s = c_0 v_m + K\left[\frac{gD(\gamma_L - \gamma_g)}{\gamma_L}\right]^{0.5}$$

where $C_0 = 1.3$ and $K = 0$ for $\theta = 0°$

$$v_s = 1.3(11.36) + 0 = 14.768\,\text{ft./s}$$

15. Liquid holdup H_L, Equation 2.37,

$$H_L = 1 - \frac{v_{sg} + (1 - E_{Ls})(v_s - v_m)}{v_s}$$

$$= 1 - \frac{5.48 + (1 - 0.781)(14.768 - 11.36)}{14.768} = 0.578$$

16. For $\theta = 0°$, calculate average liquid velocity v_L

$$v_L = \frac{v_{sL}}{H_L} = \frac{5.88}{0.578} = 10.17\,\text{ft./s}$$

17. Calculate R_{eL} and f_{wL} from Equation 2.39b

$$R_{eL} = \frac{\gamma_L D v_L}{\mu_{Lg}} = \frac{53.65(9/12)(10.17)}{12.96(32.2)(2.0886 \times 10^{-5})} = 205{,}565 > 1500$$

$$f_{wL} = \frac{0.046}{R_{eL}^{0.2}} = \frac{0.046}{(205{,}564)^{0.2}} = 3.983 \times 10^{-3}$$

18. Average slug density γ_{Ls} from Equation 2.39c

$$\gamma_{Ls} = E_{Ls}\gamma_L + (1 - E_{Ls})\gamma_g = 0.781(53.65) + (1 - 0.7)(10.95)$$

$$= 44.3\,\text{lb/ft.}^3$$

19. The fractional pressure gradient from Equation 2.39a

$$\left(\frac{dp}{dL}\right)_f = -\left(\frac{2f_L\gamma_{Ls}v_L^2}{Dg}\right) = -\frac{2(3.983 \times 10^{-3})(44.3)(10.17)^2}{(9/12)(32.2)}$$

$$= -1.51132\,\text{lb/ft.}^3$$

20. The total pressure gradient, since $\theta = 0°$

$$\left(\frac{dp}{dL}\right)_{gravitational} = 0$$

and

$$\left(\frac{dp}{dL}\right)_{total} = \left(\frac{dp}{dL}\right)_f = 1.51132\,\text{lb/ft.}^3$$

or

$$\left(\frac{dp}{dL}\right)_{total} = -1.0495 \times 10^{-2}\,\text{psi/ft.}$$

2.11 EMPIRICAL METHODS

Many empirical correlations have been developed for predicting two-phase flowing pressure gradients, which differ in the manner used to calculate three components of the total pressure gradient (see Equation 2.6). Some of these correlations are described below. The range of applicability of these multiphase flow models is dependent on several factors such as tubing size or diameter, oil gravity, gas–liquid ratio and, two-phase flow with or without water-cut. The effect of each of these factors on estimating the pressure profile in a well is discussed for all empirical methods considered [6, 7]. A reasonably good performance of the multiphase flow models, within the context of this section, is considered to have a relative error (between the measured and predicted values of the pressure profile) less than or equal to 20%.

2.12 THE DUNS-ROS METHOD [20, 21]

To better understand the initial concept of the Duns-Ros method, Figure 2.25 shows a generalized flow diagram. The Dun-Ros correlation is developed for vertical flow of gas and liquid mixtures in wells. This correlation is valid for a wide range of oil and gas mixtures with varying water-cuts and flow regimes. Although the correlation is intended for use with "dry" oil/gas mixtures, it can also be applicable to wet mixtures with a suitable correction. For water contents less than 10%, the Duns-Ros correlation (with a correction factor) has been reported to work well in the bubble, slug (plug), and froth regions. Figure 2.26 shows that pressure gradient and holdup also depend significantly on superficial gas velocity.

At low gas flowrates, the pipe essentially is full of liquid since the gas bubbles are small. Holdup is approximately equal to unity. At liquid rates less than 1.3 ft/s (0.4 m/s), increased gas rate causes the number and size of the bubbles to increase. Ultimately, they combine into plugs that become unstable and collapse at still higher gas concentrations to form slugs. At gas rates greater than 49 ft/s (15 m/s), with the same liquid rate, mist flow is initiated, and gas is in the continuous phase with liquid drops dispersed in it. When the liquid velocity is over 5.3 ft/s (16 m/s), the flow patterns are not as observable. As gas flow increases, no plug flow is observed; flow is turbulent and frothy until some degree of segregation takes place at higher rates. For this degree of liquid loading, mist flow does not occur until gas velocity reaches at least 164 ft/s (50 m/s).

Figure 2.27 graphically outlines the flow regime areas. Duns and Ros mathematically defined these areas as functions of the following

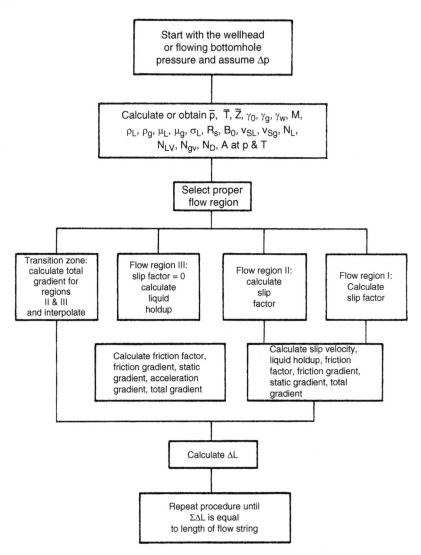

FIGURE 2.25   Flow diagram for the Duns-Ros method [1].

dimensionless numbers:

$$N_{vg} = v_{sg}A(\gamma_L/\sigma)^{0.25} \qquad \text{gas velocity number} \qquad (2.46)$$

$$N_{vL} = v_{sL}A(\gamma_L/\sigma)^{0.25} \qquad \text{liquid velocity number} \qquad (2.47)$$

$$N_d = dB(\gamma_L/\sigma)^{0.5} \qquad \text{diameter number} \qquad (2.48)$$

$$N_{L\mu} = N_L = \mu_L C(1/\gamma_L\sigma^3)^{0.25} \quad \text{liquid velocity number} \qquad (2.49)$$

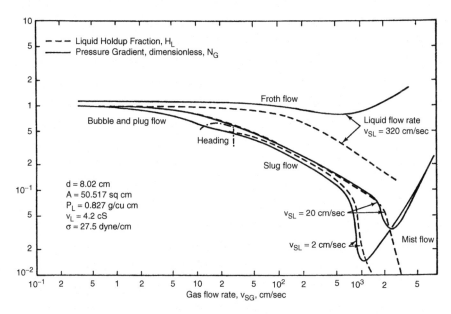

FIGURE 2.26   Example of two-phase flow in vertical pipe [5].

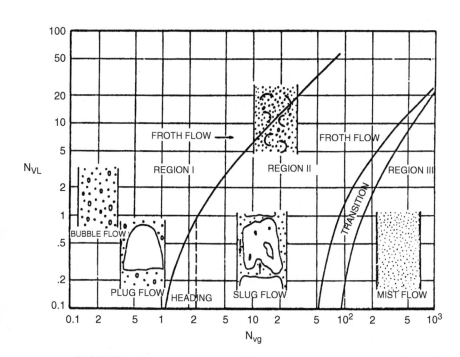

FIGURE 2.27   Region of occurrence of different flow regimes [5].

Any consistent units system may be used, for example,

	English	Metric
ρ_L (liquid density)	slugs/ft.^{3}	kg/m^3
d (diameter of pipe)	ft	M
σ (liquid surface tension)	dyn/cm	dyn/cm
μ_L (liquid velocity)	cp	cp

Equations contain a "g" term that was included into the mist conversion factors A, B, and C.

	English	Metric
A	1.938	3.193
B	120.9	99.03
C	0.1573	0.3146

At high liquid rates, the pressure gradient varied significantly with the gas rate. The various flow regions were divided into three main regions depending on the amount of gas present.

- Region I. The liquid phase is continuous and bubble flow, plug flow, and part of the froth-flow regime exists.
- Region II. In this region the phases of liquid and gas alternate. The region thus covers slug flow and the remainder of the froth flow regime.
- Region III. The gas is in a continuous phase and the mist-flow regime exists.

The different nature of these three main regions necessitates separate correlations for friction and holdup for each region; therefore in principle, six different correlations are to be expected. The identification of flow region is a function of N_{LV}, N_{gv}, L_1, L_2, and N_d. The regions of validity of the correlations are plotted and presented in Figure 2.28 as a function of the liquid velocity number N_{Lv} and gas-velocity number N_{gv}. Because N_{Lv} and N_{gv} are directly related to liquid flowrate and gas flowrate, respectively, it can be seen from Figure 2.28 that a change in one or both of these rates affects the region of flow.

Duns and Ros suggested the following limits for various flow regions:

- Region I: $0 \leq N_{gv} \leq (L_1 + L_2 N_{Lv})$
- Region II: $(L_1 + L_2 N_{Lv}) < N_{gv}, (50 + 36 N_{Lv})$
- Region III: $N_{gv} > (75 + 84 N_{Lv}^{0.75})$

L_1 and L_2 are functions of N_d, and their relationships are presented in Figure 2.29.

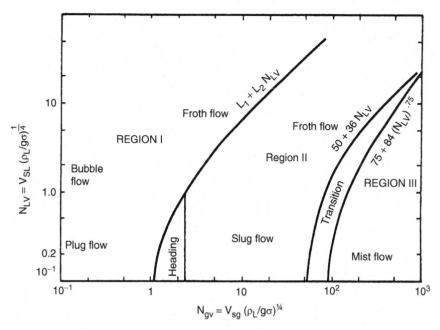

FIGURE 2.28   Region of validity of Duns-Ros correlation [5].

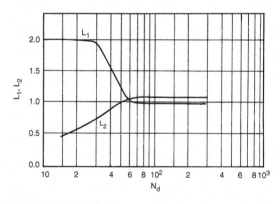

FIGURE 2.29   L factors versus diameter number, N_d [5].

It was also found that the liquid holdup is related to the slip velocity, v_s, as follows:

$$v_s = v_{sg}/(1 - H_L) - v_{sL}/H_L \qquad (2.50)$$

where v_{sg}, v_{sL} are average gas and average liquid superficial velocities, respectively.

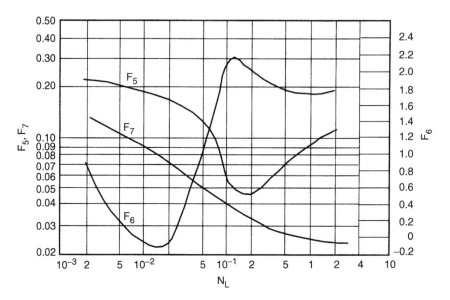

FIGURE 2.30 F_5, F_6, F_7 against viscosity number, N_L [5].

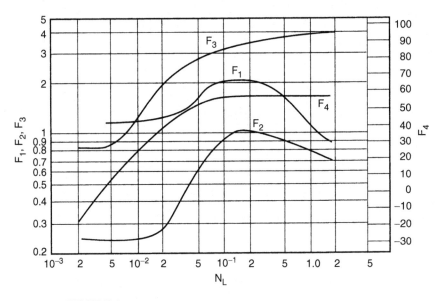

FIGURE 2.31 F_1, F_2, F_3, and F_4 versus viscosity number, N_L [5].

The slip velocity was expressed in dimensionless form as

$$S = v_s(\gamma_L/g\sigma)^{0.25}$$

As soon as S has been determined, v_s, H_L, and, finally, $(dp/dL)_{st}$ can be determined.

Different formulas are used for calculating S in each of the three flow regions. These formulas, which are functions of the four dimensionless numbers, N_{Lv}, N_{gv}, N_d, N_L, are found in the example below and make use of Figures 2.30 and 2.31.

Example 2.2

Show stepwise procedure for calculation of the pressure traverse by the Duns-Ros method. Apply this procedure to solve the following problem.

Determine the distance ΔL between two pressure points starting surface conditions if $\Delta p = 500\,psig$.

Given that tubing size $d = 2\,in. = 1.995\,in.$ ID
wellhead pressure $1455\,psig = p_1$
$p_L = p_1 + 500 = 1955\,psig$
$p_{bar} = 14.7\,psia$
$T_1 = 75°F$
$T_2 = 105°F$
$S.G._g = 0.752$
$S.G._0 = 54°API$
$q_{oil} = 480\,bpd$ (std) $q_{water} = 0\,bpd$ (std)
$q_L = q_w + q_0$
$\mu_g = 0.020\,cp$ (constant value)
$\sigma_0 = 28\,dyn/cm$
$GLR = GOR = 3393\,scf/bbl$
Choose a commercial steel pipe.

Solutions

1. Determine the specific gravity of the oil:

$$SG_0 = \frac{141.5}{131.5 + °API} = \frac{141.5}{131.5 + 54} = 0.763$$

2. Find the weight associated with 1 bbl of stock-tank liquid:

$$m = SG_0(350)\left(\frac{1}{1+WOR}\right) + SG_0(350)\left(\frac{WOR}{1+WOR}\right)$$
$$+ (0.0764)(GLR)(SG_g)$$
$$= 0.763(350)\left(\frac{1}{1+0}\right) + 0.763(350)\left(\frac{1}{1+0}\right)$$
$$+ 0.0764(3393)(0.752)$$
$$= 462\,lb/stb\ of\ oil$$

3. Determine the specific weight of the liquid phase:

$$\gamma_L = 62.4 \left[SG_0 \left(\frac{1}{1+WOR} \right) + SG_w \left(\frac{WOR}{1+WOR} \right) \right]$$

$$= 62.4 \left[0.763 \left(\frac{1}{1+0} \right) + SG_w \left(\frac{1}{1+0} \right) \right]$$

$$= 47.6 \, lb/ft.^3$$

4. Find the average pressure:

$$P_{avg} = \frac{P_1 + P_2}{2} + P_{bar} = \frac{1455 + 1955}{2} + 14.7 = 1719.7 \, psia$$

5. Find the average temperature:

$$T_{avg} = \frac{T_1 + T_2}{2} = \frac{75 + 105}{2} = 90°F = 550°R$$

6. Find Z_{avg} for gas phase

$$Z_{avg} = f(T_r, p_r)$$
$$= 0.72$$
$$T_r = T/T_{pc} = 550/394 = 1.4$$
$$p_r = 1719.7/660 = 26$$

7. Find the average specific weight of the gas phase:

$$\gamma_g = SG_g (0.0764) \left(\frac{p}{14.7} \right) \left(\frac{520}{5} \right) \left(\frac{1}{Z} \right)$$

$$= SG_g (0.0764) 1/B_g$$

$$= 0.752(0.0764) \frac{1719.7}{14.7} \frac{520}{550} \frac{1}{0.72} = 8823 \, lb/ft.^3$$

8. Find R_s at T_{avg} and p_{avg}

$$R_s = SG_g \left(\frac{p}{18} \times \frac{10^{0.0125(°API)}}{10^{0.00091(T)}} \right)^{1.2048}$$

$$T_{avg} = 90°F, \qquad p_{avg} = 1719.7 \, psia$$

$$R_s = 0.752 \left(\frac{1719.7}{28} \times \frac{10^{0.0125(54)}}{10^{0.00091(90)}} \right)^{1.2048}$$

$$= 947.3 \, scf/stb$$

9. Calculate the average viscosity of the oil from correlations

$$\mu_{od} = 10^X - 1.0$$
$$X = T^{-1.163} \exp(6.9824 - 0.04658°\text{API})$$
$$X = 90^{-1.163} \exp(6.9824 - 0.04658 \times 54) = 0.4646$$
$$\mu_{od} = 1.915\,\text{cP}$$
$$\mu_{os} = A\mu_{od}^B$$
$$A = 10.715(R_s + 100)^{-0.515} = 10.715(947.3 + 100)^{-0.515}$$
$$A = 0.2983$$
$$B = 5.44(R_s + 150)^{-0.338} = 5.44(947.3 + 150)^{-0.338}$$
$$B = 0.5105$$
$$\mu_{os} = 0.2983 \times 1.915^{0.5105} = 0.416\,\text{cp}$$

10. Determine the average water viscosity. No water is in the example.
11. Calculate the liquid mixture viscosity:

$$\mu_L = \mu_o \left(\frac{1}{1+\text{WOR}}\right) + \mu_w \left(\frac{\text{WOR}}{1+\text{WOR}}\right) = \mu_o$$
$$= 0.416\,\text{cp}$$

12. Find the liquid mixture surface tension:

$$\sigma_L = \sigma_o \left(\frac{1}{1+\text{WOR}}\right) + \mu_w \left(\frac{\text{WOR}}{1+\text{WOR}}\right) = \sigma_o$$
$$= 28\,\text{dyn/cm}$$

13. Find B_o at p_{avg} and T_{avg}.

$$B_o = 0.972 + 0.00147F^{1.175}$$
$$F = R_s \left(\frac{\gamma_g}{\gamma_o}\right)^{0.5} + 1.25\% \, [T(°F)]$$
$$= 947.3 \left(\frac{0.752}{0.763}\right)^{0.5} + 1.25(90) = 1052.9$$
$$B_o = 1.495\,\text{bbl/stb}$$

14. Find the turbine flow area A_p:

$$A_p = \frac{\pi d^2}{4} = \frac{\pi}{4} + \left(\frac{1.995}{12}\right)^2 = 0.0217\,\text{ft.}^2$$

15. Find the liquid viscosity number:

$$N_L = 0.1573 \times (0.5) \left(\frac{1}{(28)^3 47.6}\right)^{0.25} = 2.05 \times 10^{-3}$$

16. Find v_{sL} (assume $B_w = 1.0$):

$$v_{sL} = \frac{5.61qL}{86400A_p}\left[B_o\left(\frac{1}{1+WOR}\right)+B_w\left(\frac{WOR}{1+WOR}\right)\right]$$

$$= \frac{5.61(480)}{86400(0.0217)}\left[1.495(1.0)+1.0\left(\frac{0}{1+0}\right)\right]$$

$$= 2.147\,\text{ft./s.}$$

17. Find the liquid velocity number:

$$N_{Lv} = 1.938\,v_{sL}\left(\frac{\gamma_L}{\sigma_L}\right)^{1/4} = 1.938(2.147)\left(\frac{47.6}{28}\right)^{1/4}$$

$$= 4.75$$

18. Find the superficial gas velocity:

$$v_{sg} = \frac{q_L\left[GLR-R_s\left(\frac{1}{1+WOR}\right)\right]}{86,400\,A_p}\left(\frac{14.7}{p_{avg}}\right)$$

$$\times\left(\frac{T_{avg}}{520}\right)\left(\frac{Z_{avg}}{1}\right)$$

$$= \frac{q_L\left[GLR-R_s\left(\frac{1}{1+WOR}\right)\right]B_g}{86,400\,A_p}$$

$$= \frac{480\left[3393-947.3\left(\frac{1}{1+0}\right)\right]}{84,400(0.0217)}\frac{14.7\times550(0.72)}{1719.5\times520}$$

$$= 4.08\,\text{ft./s}$$

19. Find the gas velocity number:

$$N_{gv} = 1.938\,v_{sg}\left(\frac{\gamma_L}{\sigma_L}\right)^{1/4} = 1.938\times4.08\left(\frac{47.6}{28}\right)^{1/4}$$

$$= 9.03$$

20. Find the pipe diameter number:

$$N_d = 120.9d\left(\frac{\gamma_L}{\sigma_L}\right)^{0.5} = 120.9\left(\frac{1.995}{12}\right)\left(\frac{47.6}{28}\right)^{0.5}$$

$$= 26.2$$

21. Select the proper flow regime from Figure 2.3:

$$N_{gv} = 9.03$$
$$N_{Lv} = 4.75$$

These numbers fall in Region II; see Figure 2.3.

22. Determine the proper slip factor depending upon the region found in step 21.

    a. For Region I: determine the slip factor determination. The slip factor is found by the following formula:

    $$S = F_1 + F_2 N_{Lv} + F_3' \left(\frac{N_{gv}}{1 + N_{Lv}} \right)^2$$

    F_1 and F_2 are found in Figure 2.31.
    $F_3' = F_3 - F_4/N_d$ where F_3 and F_4 are found in Figure 2.31. For annular flow N_d is based on the wetted perimeter; thus, $d = (d_c + d_t)$. Region I extends from zero N_{Lv} and N_{gv} up to $N_{gv} = L_1 + L_2 N_{Lv}$, where L_1 and L_2 can be found in Figure 2.29.

    b. For Region II:

    $$S = (1 + F_5) \frac{(N_{gv})^{0.982} + F_6'}{(1 + F_7 N_{Lv})^2}$$

    F_5, F_6 and F_7 can be found in Figure 2.30 where $F_6 = 0.029 N_d + F_6$. Region II extends from the upper limit of Region I to the transition zone to mist flow given by $N_{gv} = 50 + 36 N_{Lv}$.

    c. For Region III (mist flow):

    $$S = 0$$

    $$\text{Therefore, } H_L = \frac{1}{1 + v_{sg}/v_{sL}}$$

    This is valid for $N_{gv} > 75 + 84 N_{Lv}^{0.75}$. Calculations for Region II from Figure 2.5 if $N_L = 2.05 \times 10^{-3}$, then

    $$F_5 = 0.218, \qquad F_6 = 0.58, \qquad F_7 = 0.12$$
    $$F_6' = 0.029 N_d + 0.58 = 0.029 \times (2.62) + 0.58 = 1.34$$
    $$S = (1 + 0.20) \frac{9.03^{0.982} + 1.34}{[1 + (0.12 \times 4.751)]^2} = 4.88$$

23. Determine the slip velocity if in Region I or II:

$$v_s = \frac{S}{1.938(\gamma_L/\sigma_L)^{0.5}}$$

It is Region II; hence

$$V_s = \frac{4.88}{1.938(47.61/28)^{0.5}} = 1.933$$

24. Determine the liquid holdup:

$$H_L = \frac{V_s - V_{sg} - V_{sL} + [(V_s - V_{sg} - V_{sL})^2 + 4V_s V_{sL}]^{0.5}}{2V_s}$$

$$= \left(1.933 - 4.08 - 2.147 + [(1.933 - 4.08 - 2.147)^2 \right.$$

$$\left. + 4(1.933)(2.147)]^{0.5}\right) \Big/ \left(2 \times 1.933\right)$$

$$= 0.4204 = 0.42$$

This value can be checked

$$V_s = \frac{V_{sg}}{1 - H_L} - \frac{V_{sL}}{H_L} = \frac{4.08}{1 - 0.42} - \frac{2.147}{0.42} = 1.923$$

25. Determine the liquid Reynolds number:

$$N_{RE} = \frac{1488 \gamma_L V_{sL} d}{\mu_L}$$

$$(N_{Re})_L = \frac{1488 \times 47.6 \times 2.147 \times 0.16625}{0.416} = 60,773$$

26. Determine the friction gradient according to the flow region.
    a. For Region I and II

$$G_{fr} = 2f_w \frac{N_{Lv}(N_{Lv} + N_{gv})}{N_d}$$

where $f_w = (f_1)\frac{f_2}{f_3}$

f_1 is found in Figure 2.32 and f_2 is found in Figure 2.33

The abscissa must be determined in Figure 2.33 and is $f_1 R N_d^{2/3}$

where $R = \frac{V_{sg}}{V_{sL}}$

$$f_3 = 1 + f_1(R/50)^{0.5}$$

The friction factor f_w is valid in Regions I and II and covers heading also. It is good from zero N_{Lv} and N_{gv} up to the limit given by $N_{gv} = 50 + 36 N_{Lv}$.

b. For Region III:

In mist flow where $N_{gv} > 75 + 84 N_{Lv}^{0.75}$

$$G_{fr} = 2f_w N_\rho \frac{(N_{gv})^2}{N_d}$$

where $N_\rho = \rho_g / \rho_L$

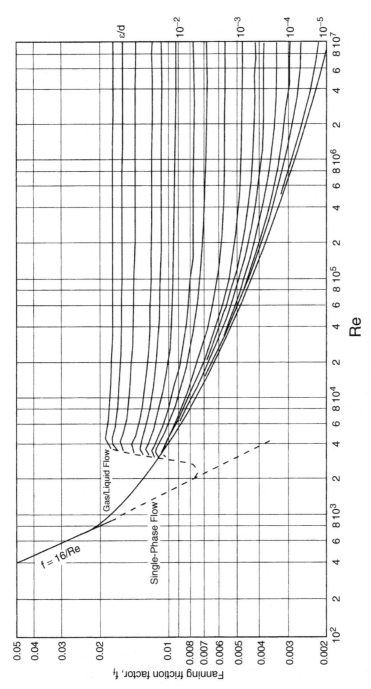

FIGURE 2.32   Fanning friction factor f_f versus Reynolds number [5].

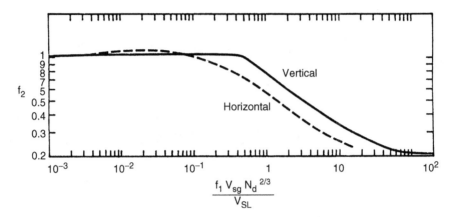

FIGURE 2.33 Bubble friction correction [5].

In Region III f_w is taken as f_1 and may be taken from Figure 2.32. For $\varepsilon > 0.05d$, the value of f_1 is calculated by

$$f_1 = \frac{1}{[4\log_{10}(0.027\varepsilon/d)]^2} + 0.067(\varepsilon/d)^{1.73}$$

For $\varepsilon > 0.05\,d$, the value of $d - \varepsilon$ should be substituted for d throughout the friction gradient calculation, and also this substitution should be made:

$$v_{sg} = \frac{v_{sg}d^2}{(d-\varepsilon)^2}$$

It is Region II; hence, calculate f_1, f_2, and f_3 such that

$$f_w = (f_1)\frac{f_2}{f_3}$$

From Figure 2.32 read f_1, but first determine a value for ε/d. If the value for ε is not known, a good value to use is 0.00015 ft, which is an average value given for commercial sted.

$$\frac{\varepsilon}{d} = \frac{0.00015(12)}{1.995} = 9.02 \times 10^{-4}$$

For given Reynolds number (60,773) and ε/d

$$f_1 = 0.005$$

From Figure 2.33

$$\text{where } f_1 \left(\frac{v_{sg}}{v_{sL}}\right) N_d^{2/3} = 0.005 \left(\frac{4.08}{2.147}\right) 26.2^{2/3}$$

$$= 0.0839$$

$$f_2 = 1.01$$

$$f_3 = 1 + f_1 \left(\frac{4.08}{2.147 \times 50} \right)^{0.5} = 1 + 0.005(0.038)^{0.5}$$

$$= 1.001$$

$$f_w = 0.005 \left(\frac{1.01}{1.001} \right) = 0.00505$$

calculate friction gradient G_{fr}:

$$G_{fr} = 2(0.00505) \frac{4.75(4.75 + 9.03)}{26.2} = 0.0252$$

27. Determine the static gradient:

$$G_{st} = H_L + (1 - H_L) \frac{\gamma_g}{\gamma_L}$$

$$= 0.42 + (1 - 0.42) \frac{8.823}{47.6} = 0.5275 \text{ (dimensionless)}$$

28. Determine the total pressure gradient.
   a. For Regions I and II:

$$G = G_{st} + G_{fr}$$

   b. For Region III (accounting for accelerations):

$$G = \frac{G_{st} + G_{fr}}{1 - (\gamma_L v_{sL} + \gamma_g v_{sg})(v_{sg}/\gamma)}$$

$$= G_{st} + G_{fr} = 0.5275 + 0.0252 = 0.5527$$

29. Convert to gradient in psi/ft:

$$\frac{dp}{dL_{st}} = \frac{G_{st}\gamma_L}{144}, \quad \frac{dp}{dL_{(fr)}} = \frac{G_{fr}\gamma_L}{144}, \quad \text{or} \quad \frac{dp}{dL_{total}} = \frac{G\gamma_L}{144}$$

$$\frac{dp}{dL} = \frac{0.5527(47.6)}{144} = 0.1827 \text{ psi/ft.}$$

30. Determine distance ΔL:

$$\Delta L = \frac{(1955 - 1455)\text{psi}}{0.1827 \text{ psi/ft.}} = 2737 \text{ ft.}$$

If water flows together with oil, it is recommended that the calculations be made using the average oil–water mixture properties.

The pressure profile prediction performance of the Duns-Ros method is outlined below in relation to the several flow variables considered [6]:

- *Tubing Size*: In general, the pressure drop is seen to be over predicted for a range of tubing diameters between 1 and 3 in.
- *Oil Gravity*: Good predictions of the pressure profile are obtained for a broad range of oil gravities (13 to 56 °API).
- *Gas-Liquid Ratio (GLR)*: The pressure drop is over predicted for a wide range of GLR. The errors become especially large (>20%) for GLR greater than 5000.
- *Water-Cut*: The Duns-Ros model is not applicable for multiphase flow mixtures of oil, water, and gas. However, the correlation can be used with a suitable correction factor as mentioned earlier.

2.13   THE ORKISZEWSKI METHOD [8, 9]

This method is recognized for four types of flow pattern, and separate correlations are prepared to establish the slippage velocity and friction for each. This correlation is limited to two-phase pressure drops in a vertical pipe and is an extension of Griffith and Wallis [10] work. The correlation is valid for different flow regimes such as the bubble, slug, transition, and annular mist and is a composite of several methods as shown below:

Method	Flow Regime
Griffith	Bubble
Griffith & Wallis	Slug (density term)
Orkiszewski	Slug (friction term)
Duns & Ros	Transition
Duns & Ros	Annular Mist

It should be noted that the liquid distribution coefficient (hold-up) is evaluated using the data from the Hagedorn and Brown model (discussed later). The correlation is applicable to high-velocity flow range and gas condensate wells in addition to oil wells and has proven accuracy. To make calculations, a computer is preferable, Figure 2.34 shows a generalized flow diagram of this method. After assuming a pressure difference and calculating the various required properties, a flow region is selected. Depending on the flow region, the pressure loss calculations—which, in general, include friction and holdup—are made. The vertical length corresponding to the

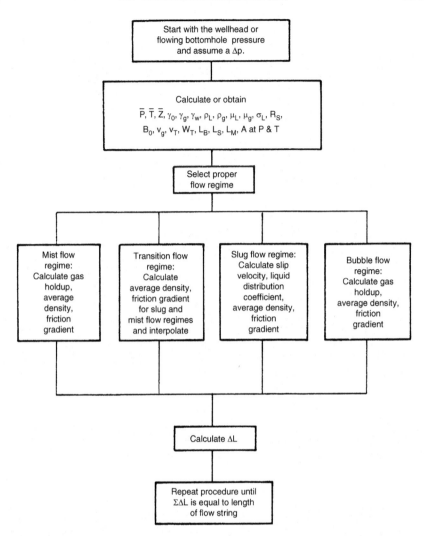

FIGURE 2.34 Flow diagram for the Orkiszewski method [1, 6].

pressure difference is then determined. The flow regime is found by testing the following limits:

Regime	Limits of Boundary Lines, L
Bubble	$q_g/(q_g + q_l) < L_B$ or $v_{sg}/v_m < L_B$
Slug	$q_g/(q_g + q_l) > L_B$, $N_{vg} < L_s$
Transition	$L_m > N_{vg} > L_s$
Mist	$N_{vg} > L_m$

The foregoing new variables are defined as follows:
Bubble:

$$L_b = 1.071 - (0.2218v_m^2/d) \quad \text{but} \quad \geq 0.13 \tag{2.51}$$

Slug:

$$L_s = 50 + 36N_{vg}q_L/q_g (\text{or } 36N_{vg}v_{sl}/v_{sg}) \tag{2.52}$$

Mist:

$$L_m = 75 + 84(N_{vg}q_L/q_g)^{0.75} \left[\text{or } 84(N_{vg}v_{sl}/v_{sg})^{0.75} \right]$$

where v_t = total fluid velocity ($v_{sl} + v_{sg} = v_m$)
$\quad q_t$ = volumetric total flow ($q_L + q_g$)
$\quad N_{vg}$ = dimensionless velocity influence number
$\quad\quad = v_{sg}(\gamma_L/g\sigma)^{0.25}$

2.13.1 Bubble Flow

The γ_m required knowledge of the holdup H_1, such that

$$\gamma_m = \gamma_L H_L + \gamma_g(1 - H_L)$$

In this, the H_L is calculated as follows:

$$H_1 = 1 - 0.5\left[1 + v_m/v_s - (1 + v_m/v_s)^2 - 4v_{sg}/v_s)^{0.5} \right] \tag{2.53}$$

where v_s = slip velocity = 0.8 ft./s (0.244 m/s)
Therefore, the friction gradient is

$$dp/dL = 2f\gamma_L v_L^2/(H_L g_c d) \tag{2.54}$$

where f = Fanning friction factor obtained from Figure 2.32

$$R_e = dv_{sL}\gamma_L/\mu_L \tag{2.55}$$

The elevation gradient is

$$dp/dL = \gamma_m F_e \tag{2.56}$$

and the acceleration gradient is negligible. However, Orkiszewski's equation for all these effects is

$$\frac{dp}{dL} = \frac{\gamma_m F_e + 2f\gamma_L v_{sL}^2/(H_L g_c d)}{1 - m_t q_g/(A_p^2 p_{avg} g_c)} \tag{2.57}$$

which is essentially the same as adding the three gradients.

2.13.2 Slug Flow

Slug flow specific weight γ_s is difficult to know and difficult to assume. An attempt was made such that

$$\gamma_s = \frac{m_t + \gamma_L v_s A_p}{q_t + v_s A_p} + \delta\gamma_L \tag{2.58a}$$

in a slightly different term arrangement using velocities, or

$$\gamma_s = \frac{\gamma_L(v_{sL} + v_s) + \gamma_g v_{sg}}{v_m + v_s} + \delta\gamma_L \tag{2.58b}$$

where $m_t =$ total mass/s

$A_p =$ area of pipe

$v_s =$ correlation factor, $C_1 C_2 (gd)^{0.5}$, slip velocity

$\delta =$ liquid distribution coefficient $\tag{2.59}$

C_1 and C_2 are functions of a Reynolds number as follows:

$$C_1 \propto (f)dv_s\gamma_L/\mu_L, \text{ or } Re_b \text{ (Figure 2.35)}$$

and

$$C_2 \propto (f)Re_b \quad \text{and} \quad Re_n = dv_t\gamma_L/\mu_L \text{ (Figure 2.36)}$$

Since v_s is a dependent variable, it must be found by iteration. A value of v_s is assumed, Re_b is calculated, and C_1 and C_2 are determined. If the calculated value of v_s does not agree with the assumed value, try again.

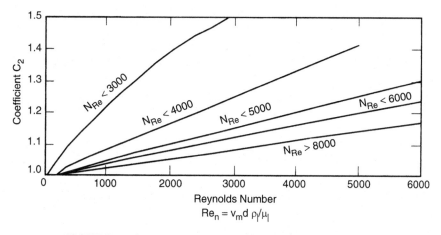

FIGURE 2.35 C_2 constant versus bubble Reynold's number [7].

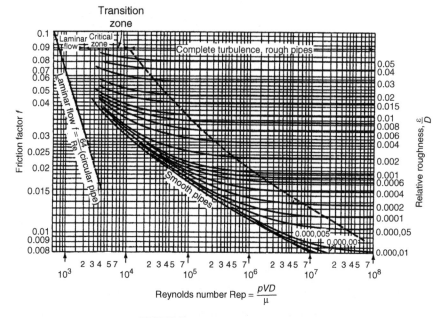

FIGURE 2.36   Friction factor [8].

[A good initial try, $v_s = 0.5(gd)^{0.5}$.] For details see Example 2.3. Now determine δ as follows:

a. If $v_m < 10$ (continuous phase is water)

$$\delta = \left[0.013 \log \mu_L/d^{1.38}\right] - 0.681 + 0.232 \log v_m$$
$$- 0.428 \log d \qquad (2.60a)$$

b. If $v_m > 10$ (continuous phase is oil)

$$\delta = \left[0.0127 \log(\mu_L + 1)/d^{1.415}\right] - 0.284 + 0.167 \log v_m$$
$$+ 0.113 \log d \qquad (2.60b)$$

c. If $v_m > 10$ (continuous phase is water)

$$\delta = \left[(0.045 \log \mu_L)/d^{0.799}\right] - 0.709 - 0.162 \log v_m$$
$$- 0.888 \log d \qquad (2.60c)$$

d. If $v_m > 10$ (continuous phase is oil) (μ_L in cp)

$$\delta = \left[0.0274 \log(\mu_L + 1)/d^{1.371}\right] + 0.161 + 0.569 \log d + x \qquad (2.60d)$$

where

$$x = -\log v_m \left[(0.01 \log(\mu_L + 1)/d^{1.571}) + 0.397 + 0.63 \log d\right]$$

These constraints apply to δ:

$$\text{If } v_m < 10, \delta \geq -0.065 \, v_m$$
$$\text{If } v_m > 10, \delta \geq (-v_x)(1 - \gamma_s/\gamma_L)/(v_m + v_x)$$

Finally, the friction gradient term is

$$\left(\frac{dp}{dL}\right)_f = \frac{2f_f\gamma_L v_m^2}{gd}\left(\frac{v_{sL} + v_s}{v_m + v_s} + \delta\right) \tag{2.61}$$

where f_f is obtained from Figure 2.32 using the following:

$$Re = dv_m\gamma_L/\mu_L \tag{2.62}$$

Again, the total pressure gradient includes the elevation (static), friction, and acceleration (negligible) components.

2.13.3 Transition Flow

Orkiszewski used linear interpolation between slug and mist.

2.13.4 Mist-Flow

The Duns-Ros method is used.

Example 2.3

Apply Orkiszewski method to solve Example 2.2 with the pipe diameter $d = 2$ in.

1. Select the starting point as the 1455 psig pressure.
2. The temperature at each point of pressure is given as 75°F at 1455 psig and 105°F at 1955 psig.
3. Δp is equal to 500 psig
4. It is not necessary to assume depth increment since the temperature at 1955 psia is known.
5. The average temperature of the increment is $(105 + 75)/2 = 90°F$
6. The following calculations are made in order to complete step 7 to determine flow regime.

   a. The average flow conditions are

$$\Delta p = (1455 + 1955)/2 + 14.7 = 1719.5$$
$$T = 90°F = 550°R$$
$$Z = 0.72$$
$$R_s = 947.3 \, \text{scf/stb}$$
$$B_0 = 1.495 \, \text{bbl/stb}$$

b. The corrected volumetric flowrates are

$$q_L = 480\,bpd = 6.4984 \times 10^{-5} \times 480\,scf/s$$
$$= 6.4984 \times 10^{-5}(q_0 B_0 + q_w B_w)$$
$$= 6.4984 \times 10^{-5} \times 480 \times 1.495$$
$$= 0.0466\,ft.^3/s$$
$$q_g = 3.27 \times 10^{-7} \times (GLR - R_s)q_L(T+460)/p$$
$$= 3.27 \times 10^{-7}(0.72)(3397 - 947.3)$$
$$\times 480(550)/1719.5$$
$$= 0.08855\,ft.^3/s$$
$$q_t = q_g + q_1 = 0.08855 + 0.0466 = 0.1352\,ft.^3/s$$

c. The corrected weight flowrates are

$$\dot{w}_L = 4.05(10^{-3})(q_0 SG_0 + q_w SG_w) + 8.85$$
$$\times (10^{-7})q_L SG_g(R_s)$$
$$= 4.05(10^{-3})(480 \times 0.763 + 0) + 8.85(10^7)480$$
$$\times (0.752)947.3$$
$$= 1.483 + 0.303 = 1.7856\,lb/s$$
$$\dot{w}_g = 8.85(10^{-7})q_L g_g(GLR - R_s)$$
$$= 8.85(10^{-7})480(0.752)(3393 - 947.3)$$
$$= 0.7813\,lb/s$$
$$\dot{w}_t = \dot{w}_L + \dot{w}_g = 1.7856 + 0.7813$$
$$= 2.567\,lb/s$$

d. The corrected specific weights are

$$\gamma_L = \dot{w}_L/q_L = 1.7856/0.0466 = 38.32\,lb/ft.^3$$
$$\gamma_g = \dot{w}_g/q_g = 0.7813/0.08855 = 8.82\,lb/ft.^3$$

7. Determine the type of flow regime:
   a. Test variables

$$A_p = 0.0217\,ft.^2$$
$$v_m = v_t = \frac{q_t}{A_p} = \frac{0.1352}{0.0217} = 6.23\,ft./s$$
$$\frac{q_g}{q_t} = \frac{0.08855}{0.1352} = 0.655$$

$$N_{vg} = 1.938\, v_{sg} \left(\frac{\gamma_L}{\sigma_L}\right)^{0.25}$$

$$v_{sg} = \frac{q_g}{A_p} = \frac{0.08855}{0.0217} = 4.081$$

$$N_{vg} = 1.938(4.081)\left(\frac{38.32}{28}\right)^{0.25} = 8.55$$

b. Boundary limits

$$d = 2/12 = 0.1662\,\text{ft.}$$

From Equation 2.51

$$\begin{aligned}
L_B &= 1.071 - (0.2218\,v_m^2/d) \\
&= 1.071 - (0.2218 \times 6.23^2/0.1662) \\
&= -50.7
\end{aligned}$$

Because L_B has such a low value, we must use $L_B = 0.13$ from Equation 2.53 $L_s = 50 + 36 N_{vg} q_L/q_g$.

$$L_s = 50 + 36(8.55)0.0466/0.09955 = 212$$

These two values, L_B and L_s, indicate that the regime is slug flow.

8. Determine the average density and the friction loss gradient
   a. Slip velocity, v_s

$$\begin{aligned}
Re_n &= dv_m\gamma_L/\mu_L = \big((0.1662)(6.23)(38.32)\big)/\big(0.5 \times 0.000672\big) \\
&= 118{,}080
\end{aligned}$$

Since this value exceeds limits of the graph (Figure 2.36) v_s must be calculated using the extrapolation equation.

$$\begin{aligned}
v_s &= 0.5(gd)^{0.5} = 0.5(32.2 \times 0.1662)^{0.5} \\
&= 1.155 \ (\text{first try}) \\
Re_b &= (1.155)(38.32)(0.1662)/(0.5 \times 0.000672) \\
&= 21{,}892
\end{aligned}$$

C_1 cannot be read from graph (Figure 2.37). To solve this problem, Orkiszewski proposed the following equations:
if $Re_b \le 3000$

$$v_s = (0.546 + 8.74 \times 10^{-6} \times Re_n)(g_cd)^{0.5}$$

if $Re_b \ge 8000$

$$v_s = (0.35 + 8.74 \times 10^{-6} Re_n)(g_cd)^{0.5}$$

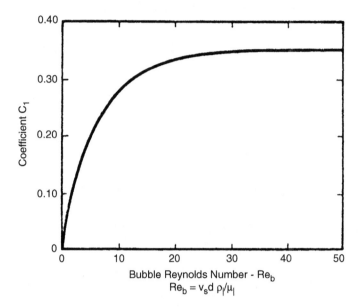

FIGURE 2.37 C_1 constant versus bubble Reynolds number [7].

if $3000 < Re_b < 8000$

$$v_s = 0.5F + \left[F^2 + 13.59\mu_L/(\gamma_L d^{0.5})\right]$$

where $F = (0.251 + 8.74 \times 10^{-6} Re_n)(g_c d)^{0.5}$
    In this example

$$v_s = 0.35 + 8.74 \times 10^{-6} \times 118{,}080 \times (32.2 \times 0.1662)^{0.5}$$
$$= 3.19\,\text{ft./s}$$

b. Liquid distribution coefficient δ and friction factor f Equation 2.60 is used to evaluate δ since $v_t < 10$ and there is no water.

$$\delta = 0.0127 \log(0.5 + 1)/0.1662^{1.415} - 0.284$$
$$+ 0.167 \log 6.23 + 0.113 \log(0.1662)$$
$$= -0.211$$

Checking this value

$$-0.211 \geq -0.065\,(6.23)$$
$$-0.211 \geq -0.405$$

Therefore, δ is okay

$$\varepsilon/d = 0.00015/0.166 = 0.00009$$

and $Re_b = 21892$

$$f_f = 0.0065 \quad \text{(Fanning factor from Figure 2.32)}$$
$$\text{or } f = 0.025 \quad \text{(friction factor from Figure 2.37)}$$

because $f = 4f_f$, so results are consistent.

c. Evaluation of average flowing specific weight from Equation 2.58

$$\gamma_s = \frac{\dot{W}_t + \gamma_L v_s A_p}{q_t + v_s A_p} + \delta \gamma_L$$
$$= \frac{2.57 + 38.32 \times 3.19 \times 0.0217}{0.1352 + (3.19)(0.0217)} + (-0.211)(38.32)$$
$$= 17.45 \, lb_m/ft.^3$$

d. Wall friction loss from Equation 2.61

$$\left(\frac{dp}{dL}\right)_f = \frac{2f_f \gamma_L v_m^2}{gd}\left(\frac{v_{sL} + v_s}{v_m + v_s} + \delta\right)$$

$$v_{sL} = 2.147 \text{ (see Example 2.2)}$$

$$\left(\frac{dp}{dL}\right)_f = \frac{2(0.0063)(38.32)(6.23)^2}{(32.2)(0.1662)} \times \left[\frac{2.174 + 3.19}{6.23 + 3.19} + (-0.211)\right]$$

$$\left(\frac{dp}{dl}\right)_f = 1.255\frac{lb}{ft.^3} = 0.0087 \, psi/ft.$$

$$\left(\frac{dp}{dL}\right)_{total} = \left(\frac{dp}{dL}\right)_{el} + \left(\frac{dp}{dL}\right)_f + \left(\frac{dp}{dL}\right)_{accl}$$
$$\Downarrow$$
$$\text{neglected}$$

$$\left(\frac{dp}{dL}\right)_{el} = \gamma_s\left(\frac{g}{g_c}\right) = 17.45\frac{32.2}{32.2}\left(\frac{lb_m}{ft.^3}\frac{ft.}{s^2}\frac{lb_f s^2}{lb_m ft.}\right)$$
$$= 17.45\frac{lb}{ft.^3} = 17.45\left(\frac{lb_f}{144\,in.}\right)\frac{1}{ft.}$$

$$\left(\frac{dp}{dL}\right)_{el} = 0.1212 \, psi/ft.$$

$$\left(\frac{dp}{dL}\right)_{total} = 0.1212 + 0.0087 = \underline{\underline{0.1299 \, psi/ft.}}$$

if $\Delta p = 500 \, psi$ since $\Delta L = 500/0.1299 = 3849 \, ft.$

The performance of Orkiszewski correlation is briefly outlined below for the flow variables considered [6].

- *Tubing Size*: The correlation performs well between 1- and 2-in. tubing sizes. The pressure loss is over predicted for tubing sizes greater than 2 in.
- *Oil Gravity*: At low oil gravities (13–30 °API), the correlation over predicts the pressure profile. However, predictions are seen to improve as oil °API increases.
- *Gas-Liquid Ratio*: The accuracy of Orkiszewski method is very good for gas–liquid ratio up to 5000. The errors become large (>20%) for GLR above 5000.
- *Water-cut*: The correlation predicts the pressure drop with good accuracy for a wide range of water-cuts.

2.14 THE HAGEDORN-BROWN METHOD [9, 12, 13]

This correlation was developed using data obtained from a 1500-ft. vertical well. Tubing diameters ranging from 1 to 2 in. were considered in the experimental analysis along with five different fluid types, namely water and four types of oil with viscosities ranging between 10 and 110 cp (@ 80°F). The correlation developed is independent of flow patterns.

The equation for calculating pressure gradient is proposed as

$$\frac{\Delta p}{\Delta L} = \gamma_{hb}(F_e) + \frac{f_f q_L^2 m_o^2}{Ad^5 \gamma_m} + \frac{(\gamma_{hb}\Delta[v_m^2/2g])}{\Delta L} \qquad (2.63)$$

We have one consistent set of units, where

		English	**Metric**
p	= pressure	lb/ft.^{2}	kPa
L	= length (height)	ft.	m
f_f	= Fanning friction factor	—	—
q_L	= total liquid flowrate	bbl/day	m^3/s
m_o	= total mass flowing/vol. liquid	slug/bbl	kg/m^3
d	= pipe ID	ft.	m
v_m	= avg. velocity = $v_{sL} + v_{sg}$	ft./sec	m/s
g	= conversion factor, force from mass	32.2 ft./sec^2	9.81 m/sec^2
A	= unit conversion constant	7.41 (10^{10})	8.63 (10^4)
γ_{HB}	= Hagedorn-Brown specific weight		
γ_s	= $v_L H_L + v_g(1 - H_L)$ based on pseudoholdup	lb/ft.^{3}	kg/m^3

γ_m	= avg. two-phase specific weight	$lb/ft.^3$	kg/m^3
V_{sL}	= superficial liquid velocity	$ft./s$	m/s
V_{sg}	= superficial gas velocity	$ft./s$	m/s
γ_L	= liquid specific weight	$lb/ft.^3$	kg/m^3
γ_g	= gas specific weight	$lb/ft.^3$	kg/m^3
H_L	= liquid holdup, a fraction		
F_e	= force equivalent	1.0	9.81

The friction factor used in Equation 2.63 is found from Figure 2.32. Figure 2.38 provides a relative roughness number. For this method, the Reynolds number for use with Figure 2.32 is

$$R_e = \frac{2.2(10^{-2})q_L m_o}{d\mu_m} \tag{2.64}$$

where

	English	Metric
q = volumetric flowrate total all fluids	ft^3/s	m^3/s
μ_m = averaged viscosity, using an equation of the form of Arrhenius:		
$\mu_m = \mu_L^{H_L} \mu_g^{(1-H_L)}$		
and		
$q_L \times \dot{m}_0 = \dot{w}_t$		

The equation would be solved over finite segments of pipe. Δv_m^2 is the changed velocity at points 1 and 2, the inlet to and outlet from that section. γ_m is the specific weight at the average p and T in the section.

Figure 2.39 also contains two empirical correction factors C and ψ. A plot of the data showed that holdup versus viscosity was a series of essentially straight lines. Water was chosen arbitrarily as a base curve (C = 1.0). C then is used for other viscosity fluids to make the parallel curves coincident. The viscosity correction curve obtained is shown in Figure 2.40.

The factor ψ was included to fit some of the data where it was postulated that a transition would occur before mist-flow begins, with gas velocity as the major variable. As gas velocity approached that required for mist-flow, it breaks through the liquid phase and the turbulence produces a liquid "ring," which increases slippage. As velocity increases even further, the shear forces on this ring dissipate it until the primary mechanism is mist flow. Figure 2.41 shows the correlation for ψ. In most cases ψ will be equal to 1.0.

The basic correlating equation, 2.64 can be converted to a form similar to that for either single flow by allowing $H_L \rightarrow 0$ for gas or $H_L \rightarrow 1.0$ for liquids. As gas rate or liquid rate approaches zero, the pressure gradient

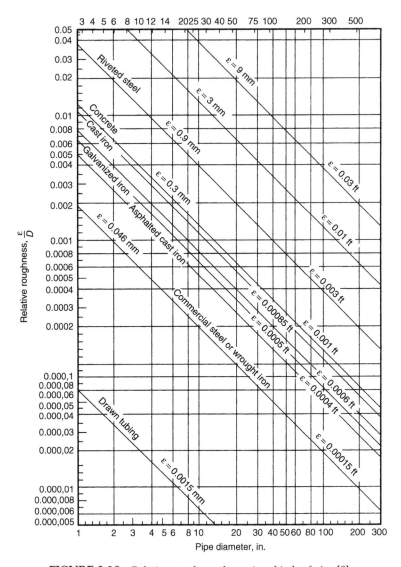

FIGURE 2.38   Relative roughness for various kinds of pipe [8].

obtained likewise approaches that for the other single phase. One therefore has a continuous gradient from liquid to two-phase to gas, an important aspect of the model.

Reviewing the foregoing calculation summary, it is necessary to make the calculation for a given diameter pipe and a given flowrate to avoid a trial-and-error solution. One can find R_e and all of the velocity-associated numbers to solve Equation 2.64. This would have to be repeated for various pipe

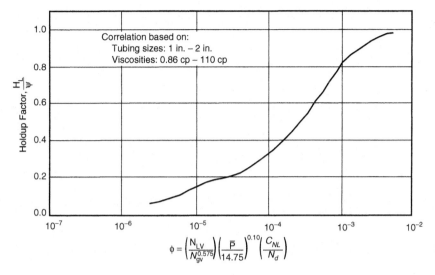

$$\phi = \left(\frac{N_{LV}}{N_{gv}^{0.575}}\right)\left(\frac{\bar{P}}{14.75}\right)^{0.10}\left(\frac{C_{NL}}{N_d}\right)$$

FIGURE 2.39 Holdup factor correlation [9].

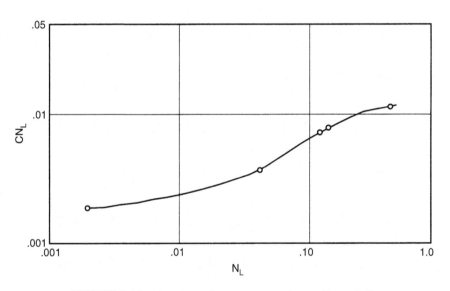

FIGURE 2.40 Correlation for viscosity number coefficient [10].

size holdup calculations based primarily on data from 1.25-in. (0.031-m) tubing, the correlation in Figure 2.39 resulted. Some of the forms in this figure are the output from characterizing numbers and secondary correlations. Four dimensionless characterizing numbers were first proposed by Ros and adapted by others. They are given by Equations 2.46 to 2.49.

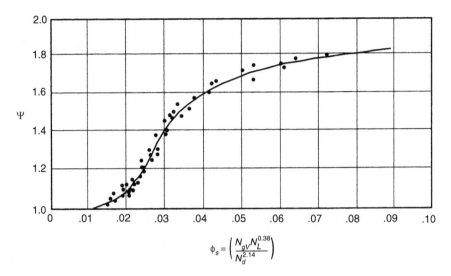

$$\phi_s = \left(\frac{N_{gv}N_L^{0.38}}{N_d^{2.14}} \right)$$

FIGURE 2.41   Correlation for secondary factor correction [10].

When the liquid stream contains both oil and water, calculate the properties as follows:
the liquid-specific weight γ_L:

$$\gamma_L = \left[\frac{SG_o(62.4) + R_sSG_g(0.0764)/5.614}{B_o} \right] \left(\frac{1}{1+WOR} \right)$$
$$+ \left[SG_w(62.4)\left(\frac{WOR}{1+WOR} \right) \right] \tag{2.65}$$

the total weight associated with 1 bbl of stock tank liquid, w:

$$w = SG_o(350)\left(\frac{1}{1+WOR} \right) + SG_w(350)\left(\frac{WOR}{1+WOR} \right)$$
$$+ (0.0764)(GLR)SG_g \tag{2.66}$$

the weight flowrate $\dot{w}_t$:

$$\dot{w}_t = \dot{w}(q_w + q_o) \tag{2.67}$$

the liquid mixture viscosity μ_L:

$$\mu_L = \mu_o\left(\frac{1}{1+WOR} \right) + \mu_w\left(\frac{WOR}{1+WOR} \right) \tag{2.68}$$

the liquid mixture surface tension σ:

$$\sigma_L = \sigma_o\left(\frac{1}{1+WOR} \right) + \sigma_w\left(\frac{1}{1+WOR} \right) \tag{2.69}$$

the superficial liquid velocity v_{sL} in ft/s (assuming $B_w = 1.0$):

$$v_{sL} = \frac{5.61\,q_L}{86,400A_p} \left[B_o \left(\frac{1}{1+WOR} \right) + B_w \left(\frac{WOR}{1+WOR} \right) \right] \qquad (2.70)$$

the superficial gas velocity v_{sg}:

$$v_{sg} = \frac{q_L \left[GLR - R_s \left(\dfrac{1}{1+WOR} \right) \right]}{86,400\,A_p} \left(\frac{14.7}{p} \right) \left(\frac{T}{520} \right) \left(\frac{Z}{1} \right) \qquad (2.71)$$

Mixture specific weight is calculated by both using the Hagedorn-Brown holdup correlation and assuming no slippage. The higher value is then used.

If bubble flow is the dominant regime, the pressure gradient is used in the same way as in the Orkiszewski approach (step 7).

Example 2.4
Solve the problem in Example 2.2 using the Hagedorn-Brown method:

$$p_1 = 1455\,psig$$
$$p_i = 1955\,psig$$
$$d = 1.995 \text{ in ID } (A_p = 0.0217)$$
$$p_{bar} = 14.7\,psia$$
$$T_1 = 75°F$$
$$T_2 = 105°F$$
$$SG_g = 0.752$$
$$SG_0 = 54°API$$
$$q_0 = 480\,bpd$$
$$q_w = 0$$
$$\mu_s = 0.020\,cp$$
$$\sigma_0 = 28\,dyn/cm$$
$$GLR = 3393\,scf/bbl$$

Solution
1. $p = (1455 + 1955)/2 + 14.7 = 1179.7$
2. $T = (75 + 105)/2 = 90°F$
3. $SG_0 = 0.763$ from Example 2.2
4. Total mass flowing from volume liquid-$m_o = 462\,lb/stb$ of oil from Example 2.2
5. Solution gas/oil ratio R_s and oil formation volume factor B_o

$$R_s = 947.3\,scf/stb$$
$$B_o = 1.495\,bbl/stb$$

6. Liquid specific weight γ_L and gas specific weight γ_g from Equation 2.65

$$\gamma_L = \frac{(0.763)62.4 + 947.3(0.752)(0.0764)}{(1.495)\qquad(5.614)} = 38.33\,\text{lb/ft.}^3$$

$$\gamma_g = 8.823\,\text{lb/ft.}^3$$

7. Oil viscosity, μ_{os} and oil surface tension, σ_0

$$\mu_{os} = 0.5\,\text{cp}$$
$$\sigma_0 = \sigma_L = 28\,\text{dyn/cm}$$

8. $N_L = 0.1573(0.5)(1/(38.33 \times 28^3))^{0.25} = 2.6 \times 10^{-3}$ from Equation 2.49
9. $CN_L = 0.0022$ (from Figure 2.40)
10. $v_{sL} = (5.61 \times 480)/(86{,}400 \times 0.0217)1.495 = 2.147\,\text{ft./s}$
    from Equation 2.70
11. $N_{vL} = 1.938(2.147)(38.33/28)^{0.25} = 4.5$ from Equation 2.65
12. $v_{sg} = 4.08\,\text{ft./s}$ from Equation 2.71
13. $N_{gv} = 1.938 \times 4.08(38.33/28)^{0.25} = 8.55$ from Equation 2.46
14. Check the flow regime; calculate A and B:

$$A = L_B = 1.071 - \left[0.2218(4.08 + 2.147)^2\right]/0.1662$$
$$= -50.68 \text{ from Equation 2.51}$$

The minimum limit for L_B is 0.13. To assume 0.13

$$B = v_{sg}/(v_{sL} + v_{sg}) = 4.08/(4.08 + 2.147) = 0.655$$

Since $B - A = 0.616 - 0.13 = 0.486$, the difference is positive; so continue the Hagedorn-Brown procedure. In case B-A is negative, use the Orkiszewski method to find flow regime.

15. $N_d = 120.9(0.1662)(38.33/28)^{0.5} = 23.5$
16. Calculate holdup correlating function ϕ

$$\phi = \frac{N_{Lv}}{N_{gv}^{0.575}} \left(\frac{p}{14.7}\right)^{0.1} \left(\frac{CN_L}{N_d}\right)$$

$$= \frac{4.5}{8.55^{0.575}} \left(\frac{1719.7}{14.7}\right)^{0.1} \left(\frac{0.0022}{23.5}\right) = 0.000197$$

17. Now from Figure 2.39, $H_L/\psi = 0.42$
18. Calculate secondary correlation factor ϕ_s

$$\phi_s = \frac{N_{gv}N_L^{0.38}}{Nd^{2.14}} = \frac{8.55 \times (2.6 \times 10^{-3})0.38}{(23.5)^{2.14}} = 0.001036$$

19. From Figure 2.41, $\psi = 1.0$
20. Liquid holdup $H_L = (H_L/\psi)(\psi) = 0.42 \times 1.0 = 0.42$

21. The two-phase Reynolds number $(Re)_{tp}$ is

$$Re_{tp} = \frac{2.2 \times 10^{-2} \times 221{,}760}{(0.1662)0.5^{0.42}0.02^{0.58}} = 379{,}760$$

22. Relative roughness $\varepsilon/d = 0.00015/0.1662 = 0.0009$
23. Fanning friction factor from Figure 2.32

$$f_f = 0.00510$$

24. Calculate the average two-phase specific weight γ_m using two methods:

   a. $\gamma_m = \gamma_{HB} = \gamma_L H_L + \gamma_g(1 - H_L)$
   $$= 38.33(0.42) + 8.823(0.58) = 21.22\,lb/ft.^3$$

   b. $\gamma_m = \dfrac{\dot{w}}{V_m}$

   $$= \frac{350SG_0 + 0.0764SG_g(GOR) + 350SG_w(WOR)}{5.61B_0 + 5.61(WOR) + (GOR - R_s)B_g}$$

   $$= \frac{350 \times 0.763 + 0.0764(0.752)3393 + 0}{5.61(1.495) + (3393 - 947.3) \times (14.7/1719.7)(550/520)(0.72)}$$

   $$= 19.006\,lb/ft^3.$$

   Use $21.22\,lb/ft^3$ as a proper value.

25. Calculate

   a. $Z_1, B_{01}, R_{s1}, v_{sL2}$ and v_{sg1} at T_2, p_1 (repeat steps 6, 7, 9, 16, 18)

   $$\left.\begin{array}{l} T_1 = 75°; p_1 - 1455\,psig \\ T_r = (75 + 460)/394 = 1.358 \\ p_r = 1455/660 = 2.2 \end{array}\right\} \quad Z_1 = 0.71$$

   $$R_{s1} = 0.752[(1455/18) \times 10^{0.0125(54)}/10^{0.00091(75)}]1.2048$$
   $$= 804.4\,scf/stb$$

   $$B_{01} = 0.972 + 0.000147\,F^{1.175} = 1.403\,bbl/stb$$

   where $F = 804.4(0.752/0.763)^{0.5} + 1.25(75) = 892.3$

   $$v_{sL1} = \left(\frac{5.61 \times 480}{86{,}400 \times 0.0217}\right)(1.403) = 2.02\,ft./s$$

   $$v_{sg1} = \frac{[480(3393 - 804.4)14.7 \times 535 \times 0.71]}{86{,}400 \times 0.0217 \times 1{,}455 \times 520}$$
   $$= 4.89\,ft./s$$

b. $Z_2, B_{02}, R_{s2}, v_{sL2},$ and v_{sg2} at T_2, p_2

$$T_2 = 105°F \qquad p_2 = 1955$$

$$\left.\begin{array}{l} T_r = (105+460)/394 = 1.434 \\ p_r = 1955/660 = 2.962 \end{array}\right\} \quad Z_2 = 0.73$$

$$R_{s2} = 0.752[(1955/18) \times 10^{0.0125(54)}/10^{0.00091(105)}]^{1.2048}$$
$$= 1064.5\,scf/stb$$

26. Calculate the two-phase velocity at both P_q and P_2:

$$V_{m1} = V_{sL1} + V_{sg1} = 2.02 + 4.89 = 6.91$$
$$V_{m2} = V_{sL2} + V_{sg2} = 2.262 + 3.368 = 5.63$$

27. Determine value $\Delta\left(V_m^2\right)$

$$\Delta\left(V_m^2\right) = (6.91^2 - 5.63^2) = 16.05$$

28. From Equation 2.64

$$\Delta L = \frac{\Delta p - \gamma_{HB}\Delta\left(\dfrac{V_m^2}{2g_c}\right)}{\gamma_{HB}(F_e) + (f_f q_L^2 m_0^2/Ad^5\gamma_m)}$$

$$\Delta p = 500\,psi = 500 \times 144\,lb/ft.^2$$

$$\gamma_{HB} = \gamma_m = 21.22, \qquad A = 7.41 \times 10^{10}$$

$$\Delta L = \frac{500(144) - 21.22\dfrac{16.05^2}{2 \times 32.2}}{21.22(1) + \dfrac{[0.0051(480)^2(462)^2]}{[7.41 \times 10^{10}(0.1662)^2 21.22]}}$$

$$= (72,000 - 84.88)/(21.22 + 0.005774) = 3388\,ft.$$

or

$$\frac{\Delta p}{\Delta L} = \frac{500}{3388} = 0.1476\,psi/ft.$$

The performance of the Hagedorn-Brown method is briefly outlined below [6].

- *Tubing Size:* The pressure loses are accurately predicted for tubing size between 1 and 1.5 in., the range in which the experimental investigation was conducted. A further increase in tubing size causes the pressure drop to be over predicted.

- *Oil Gravity:* The Hagedorn-Brown method is seen to over predict the pressure loss for heavier oils (13–25 °API) and under predict the pressure profile for lighter oils (45–56 °API)
- *Gas-Liquid Ratio:* The pressure drop is over predicted for gas–liquid ratio greater than 5000.
- *Water-cut:* The accuracy of the pressure profile predictions is generally good for a wide range of water-cuts.

2.15   THE BEGGS-BRILL METHOD [2, 9, 15]

The Beggs and Brill correlation is developed for tubing strings in inclined wells and pipelines for hilly terrain. This correlation resulted from experiments using air and water as test fluids over a wide range of parameters given below:

gas flowrate 0 to 300 Mscfd
liquid flowrate 0 to 30 gal/min
average system pressure 35 to 95 psia
pipe diameter 1 and 1.5 in.
liquid holdup 0 to 0.870
pressure gradient 0 to 0.8 psi/ft.
inclination angle −90° to +90° also horizontal flow patterns

A flow diagram for calculating a pressure traverse in a vertical well is shown in Figure 2.42. The depth increment equation for ΔL is

$$\Delta L = \frac{\Delta p \left(1 - \frac{\gamma_t v_t v_{sg}}{gp}\right)}{\gamma_t \sin\theta + \frac{f_t G_t v_t}{2gd}} \tag{2.72}$$

where γ_t = two-phase specific weight in $lb/ft.^3$
$\quad v_t$ = two-phase superficial velocity
$\quad\quad (v_t = v_{sL} + v_{sg})$ in ft./s
$\quad f_t$ = two-phase friction factor
$\quad G_t$ = two-phase weight flux rate $(lb/s \cdot ft.^2)$

A detailed procedure for the calculation of a pressure traverse is following:

1. Calculate the average pressure and average depth between the two points:

$$p = (p_1 + p_2)/2 + 14.7$$

2. Determine the average temperature T at the average depth. This value must be known from a temperature versus depth survey.

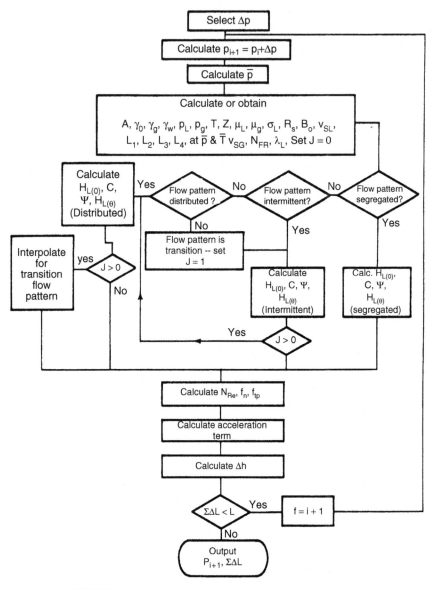

FIGURE 2.42   Flow diagram for the Beggs-Brill method [1].

3. From P–V–T analysis or appropriate correlations, calculate $R_s, B_0, B_w, \mu_0, \mu_w, \mu_h, \sigma_0, \sigma_{ws}$, and Z at T and p_s.

4. Calculate the specific gravity of the oil SG_0:

$$SG_0 = \frac{141.5}{131.5 + API}$$

5. Calculate the liquid and gas densities at the average conditions of pressure and temperatures:

$$\gamma_L = \gamma_0 \left(\frac{1}{1+WOR} \right) + \gamma_w \left(\frac{1}{1+WOR} \right) = \gamma_0 f_0 + \gamma_w f_w$$

$$\gamma_0 = \frac{350SG_0 + 0.0764R_s SG_g}{5.615}$$

$$\gamma_w = \frac{250SG_w}{5.615B_w}$$

$$\gamma_g = \frac{0.0764SG_g p(520)}{(14.7)(T+460)Z}$$

6. Calculate the *in situ* gas and liquid flowrates.

$$q_g = \frac{3.27 \times 10^{-7}Zq_0(R-R_s)(T+460)}{P}$$

$$q_L = 6.49 \times 10^{-5}(q_0 B_0 + q_w B_w)$$

7. Calculate the in situ superficial gas, liquid and mixture velocities:

$$v_{sL} = q_L/A_p$$
$$v_{sg} = q_g/A_p$$
$$v_t = v_{sL} + v_{sg}$$

8. Calculate the liquid, gas, and total weight flux rates:

$$G_L = \gamma_L v_{sL}$$
$$G_t = G_L + G_g$$
$$G_g = \gamma_s v_{sg}$$

9. Calculate the input liquid content (no-slip holdup):

$$\lambda = \frac{q_L}{q_L + q_g}$$

10. Calculate the Froude number N_{FR}, the liquid viscosity, μ_L, the mixture viscosity μ_m and the liquid surface tension σ_L:

$$N_{FR} = \frac{v_t^2}{gd}$$
$$\mu_L = \mu_0 f_0 + \mu_w f_w$$
$$\mu_t = \mu_L \lambda + \mu_g (1-\lambda)(6.72 \times 10^{-4})$$
$$\sigma_L = \sigma_0 f_0 + \sigma_w f_w$$

11. Calculate the no-slip Reynolds number and the liquid velocity number:

$$(N_{Re})_{ns} = \frac{G_t d}{\mu_t}$$

$$N_{LV} = 1.938 v_{sL} \left(\frac{\gamma_L}{\sigma_L}\right)^{0.25}$$

12. To determine the flow pattern that would exist if flow were horizontal, calculate the correlating parameters, $L_1, L_2, L_3,$ and L_4:

$$L_1 = 316\lambda^{0.0302}$$
$$L_3 = 0.10\lambda^{-1.4516}$$
$$L_2 = 0.0009252\lambda^{-2.4684}$$
$$L_4 = 0.5\lambda^{-6.738}$$

13. Determine flow pattern using the following limits:
Segregated:

$$\lambda < 0.01 \quad \text{and} \quad N_{FR} < L_1$$

or

$$\lambda \geq 0.01 \quad \text{and} \quad N_{FR} < L_2$$

Transition:

$$\lambda \geq 0.01 \quad \text{and} \quad L_2 < N_{FR} < L_3$$

Intermittent:

$$0.01 \leq \lambda < 0.4 \quad \text{and} \quad L_3 < N_{FR} < L_1$$

or

$$\lambda \geq 0.4 \quad \text{and} \quad L_3 < N_{FR} \leq L_4$$

Distributed:

$$\lambda < 0.4 \quad \text{and} \quad N_{FR} \geq L_1$$

or

$$\lambda \geq 0.4 \quad \text{and} \quad N_{FR} > L_4$$

14. Calculate the horizontal holdup $H_L(O)$:

$$H_L(O)\frac{a\lambda^b}{N_{FR}^c}$$

where a, b and c are determined for each flow pattern from the following table:

Flow pattern	a	b	c
Segregated	0.98	0.4846	0.0868
Intermittent	0.845	0.5351	0.0173
Distributed	1.065	0.5824	0.0609

15. Calculate the inclination correction factor coefficient:

$$C = (1-\lambda)\ln\left(d\lambda^e N_{LV}^f N_{FR}^g\right)$$

where d, e, f, and g are determined for each flow condition from the following table:

Flow Pattern	d	e	f	g
Segregated uphill	0.011	−3.768	3.539	−1.614
Intermittent uphill	2.96	0.305	−0.4473	0.0978
Distributed uphill	No correction		$C = 0$	
All flow patterns downhill	4.70	−0.3692	0.1244	−0.5056

16. Calculate the liquid inclination correction factor:

$$\psi = 1 + C[\sin(1.8\theta) - 0.333\sin^3(1.8\theta)] = 1 + 0.3C$$

for vertical well

17. Calculate the liquid holdup and the two-phase density:

$$H_L(\theta) = H_L(0)\psi$$
$$\rho_t = \rho_L H_L + \rho_g(1 - H_L)$$

18. Calculate the friction factor ratio:

$$f_t/f_{ns} = e^s$$

where

$$S = \frac{\ln(y)}{-0.0523 + 3.182\ln(y) - 0.8725[\ln(y)]^2 + 0.01853[\ln(y)]^4}$$

$$y = \lambda/[H_L(\theta)]^2$$

S becomes unbounded at a point in the interval $1 < y < 1.2$; and for y in this interval, the function S is calculated from

$$S = \ln(2.2y - 1.2)$$

19. Calculate the no-slip friction factor:

$$f_{ns} = 1/\{2\log[N_{Rens}/(4.5223\log N_{Rens} - 3.8215)]\}^2$$

or

$$f_{ns} = 0.0056 + 0.5/(N_{Rens})^{0.32}$$

20. Calculate the two-phase friction factor:

$$f_t = f_{ns}/(f_t/f_{ns})$$

21. Calculate ΔL. If the estimated and calculated values for ΔL are not sufficiently close, the calculated value is taken as the new estimated value and the procedure is repeated until the values agree. A new pressure increment is then chosen and the process is continued until the sum of the ΔL's is equal to the well depth.

Example 2.5
Solve the problem in Example 2.2 using the Beggs-Brill method.

Solution
1. $p = 1719.7\,\text{psia}$
2. $T = 90°F$
3. $R_s = 947.3\,\text{scf/stb}$ $B_0 = 1.495\,\text{bbl/stb}$
   $\mu_{os} = 0.5\,\text{cp}$, $\sigma_0 = 28\,\text{dyn/cm}$, $Z = 0.72$
4. $SG_0 = 0.736$, $\gamma_g = 8.823\,\text{lb/ft.}^3$
5. $\gamma_0 = 38.32\,\text{lb/ft.}^3$ (from Example 2.3)
6. $q_g = 0.08855\,\text{ft.}^3/\text{s}$
   $q_L = 0.0466\,\text{ft.}^3/\text{s}$
7. $A_p = 0.0217\,\text{ft.}^2$
   $v_{sL} = q_L/A_p = 2.147\,\text{ft./s}$, $v_{sg} = 4.081\,\text{ft./s}$
8. Calculate the liquid, gas, and total weight flux rates:

$$G_L = \gamma_L v_{sL}, \quad G_g = \gamma_g v_{sg}$$
$$G_t = G_L + G_g = 38.32 \times 2.147 + (8.823) \times 4.081$$
$$= 118.3\,\text{lb}/(\text{s·ft.}^2)$$

9. Calculate the input liquid (no-slip holdup):

$$\lambda = \frac{q_L}{q_L + q_g} = \frac{0.0466}{0.0466 + 0.8855} = 0.3448 = 0.345$$

10. The Froude number, viscosity, and surface tension

$$N_{FR} = \frac{v_t^2}{gd} = \frac{6.23^2}{32.174 \times 0.1662} = 7.26$$

$$\mu_L = \mu_o f_o + \mu_w f_w = 0.5(1.0) + \mu_w(0.0) = 0.5$$

$$\mu_t = (6.72 \times 10^{-4})[0.5 \times 0.345 + 0.02(1 - 0.345)]$$

$$= 1.164 \times 10^{-4}$$

$$= 0.0001164 \, lb_m/(ft./s)$$

$$\sigma_L = \sigma_o f_o + \sigma_w f_w = 28 \times 1.0 = 28 \, dyn/cm$$

11. Calculate the no-slip Reynolds number and the liquid velocity number:

$$(N_{Re})_{ns} = \frac{G_t d}{\mu_t} = \frac{118.3(0.1662)}{0.000464} = 168{,}884$$

$$N_{LV} = 1.938 \times 2.147(38.32/28)^{0.25} = 4.5$$

12. Determine the flow pattern that would exist if flow were horizontal:

$$L_1 = 316\lambda^{0.302} = 316 \times (0.345)^{0.302} = 229.14$$

$$L_2 = 0.0009252(0.345)^{-2.4684} = 1.2796 \times 10^{-2}$$

$$L_3 = 0.10\lambda^{-1.4516} = 0.10(0.345)^{-1.4516} = 0.4687$$

$$L_4 = 0.5\lambda^{-6.738} = 0.5(0.345)^{-6.738} = 650.3$$

13. Determine flow pattern:

$$0.4 > \lambda > 0.01 \quad \text{and} \quad L_2 < N_{FR} < L$$

The flow pattern is intermittent.

14. Calculate the horizontal holdup:

$$H_L(O) = 0.845(0.345)^{0.5351}/7.26^{0.0173} = 0.462$$

15. Calculate the inclination correction factor coefficient:

$$C = (1 - 0.345)\ln\left(2.96 \times 0.345^{0.305}4.5^{-0.4473}7.26^{0.0978}\right)$$

$$= 0.18452$$

16. Calculate the liquid holdup inclination correction factor:

$$\psi = 1 + C[\sin(1.8 \times 90) - 0.333 \sin^3(1.9 \times 90)]$$

$$= 1 + C(0.309 - 0.009826) = 1 + 0.3C$$

$$= 1 + 0.3(0.18452) = 1.055$$

17. Calculate the liquid holdup and the two-phase density:

$$H_L(90) = H_L(O)\psi = 0.462 \times 1.055 = 0.4876$$
$$\gamma_t = \gamma_L H_L + \gamma_g(1 - H_L)$$
$$= 38.32(0.4876) + 8.823(1 - 0.4876)$$
$$= 23.2\,lb/ft.^3$$

18. Calculate the friction factor ratio:

$$y = [0.345/(0.4876)^2] = 1.451, \ln 1.451 = 0.3723$$
$$f_t/f_{ns} = \exp[0.3723/(-0.0523 + 3.182 \times 0.3723)$$
$$- (0.8725 \times 0.3723^2 + 0.1853 \times 0.3723^4)]$$
$$= \exp(0.3723/1.0188) = 3^{0.36796} = 1.4447$$

19. Calculate the no-slip friction factor:

$$f_{ns} = 1/\{2\log[N_{Rens}/(4.5223 \, \log N_{Rens} - 3.8215)]\}^2$$
$$= 1/36.84 = 0.0271$$

20. Calculate the two-phase friction factor:

$$f_t = f_{ns}(f_t/f_{ns}) = 0.0271(1.4447) = 0.0391$$

21. Determine the distance ΔL for $\Delta p = 500$ psi from Equation 2.72

$$\Delta L = \frac{500\left[1 - \dfrac{23.2(6.23)4.081}{32.174 \times 1719.7}\right]144}{23.2(1.0) + \dfrac{0.0391(118.3)6.23}{2(32.174)0.1662}} = \frac{72,000(0.9893)}{25.894}$$
$$= 2750\,ft$$

and

$$\frac{\Delta p}{\Delta L} = \frac{500}{2750} = 0.18\,psi/ft.$$

Example 2.6

Solve Example 2.1 using the Beggs-Brill method:

$q_g = 40\,MMscf/d, \ p_{avg} = 2000\,psia$
$q_0 = 40,000\,stb/d, \ T_{avg} = 80°F$
$ID = 9\,in. \ SG_g = 0.75$ at $p = 14.7$ psia in $T = 60°F$
$R_p = 990\,scf/bbl$

Solution

1. $SG_0 = 141.5/131.5 + API = 0.86$
2. Calculate R_s, B_0, μ_g, Z_g at p_{avg} and T_{avg}:

$$Z_g = 0.685$$
$$\mu_g = 0.0184\,cp$$
$$R_s = 477\,scf/stb$$
$$B_0 = 1.233\,rb/stb$$
$$\mu_0 = 2.96\,cp$$

3. Calculate γ_0 and γ_g at average parameters:

$$\gamma_0 = \frac{350(0.86) + 0.0764(477)(0.75)}{5.614(1.233)} = 47.42\,lb/ft.^3$$

$$\gamma_g = \frac{0.0764\gamma_g p(520)}{(14.7)(T+460)Z_g} = \frac{0.0764(0.73)(2000)(520)}{(14.7)(80 \times 460)(0.685)}$$
$$= 10.96\,lb/ft.^3$$

4. Calculate the in situ gas and liquid flowrates,

$$q_g = \frac{3.27 \times 10^{-7} Z_g q_0 (R - R_s)(T+460)}{P}$$
$$= \frac{3.27 \times 10^{-7}(0.685)(40,000)(990 - 477)(80 + 460)}{2000}$$
$$= 1.241\,ft.^3/s$$

$$q_L = 6.49 \times 10^{-5}(q_0 B_0 + q_w B_w)$$
$$= 6.49 \times 10^{-5}[40,000(1.233) + 0]$$
$$= 3.201\,ft.^3/s$$

5. Calculate A_p:

$$A_p = \frac{\pi}{4}D^2 = \frac{\pi}{4}\left(\frac{9}{12}\right)^2 = 0.4418\,ft.^2$$

6. Calculate the in situ superficial gas, liquid, and mixture velocities:

$$v_{sL} = q_L/A_p = 3.201/0.4418 = 7.25\,ft./s$$
$$v_{sg} = q_g/A_p = 1.241/0.4418 = 2.81\,ft./s$$
$$v_m = v_{sL} + v_{sg} = 7.25 + 2.81 = 10.06\,ft./s$$

7. Calculate the liquid, gas, and total mass flux rates:

$$G_L = \rho_L v_{sL} = (47.42)(7.25) = 343.6 \, lb/(s \cdot ft.)$$
$$G_g = \rho_g v_{sg} = (10.96)(2.81) = 30.79 \, lb/(s \cdot ft.)$$
$$G_m = G_L + G_g = 343.6 + 30.8 = 374.4 \, lb/(s \cdot ft.)$$

8. Calculate the no-slip holdup:

$$\lambda = \frac{q_L}{q_L + q_s} = \frac{3.201}{32 + 1.241} = 0.72$$

9. Calculate the Froude number N_{FR}, the mixture viscosity μ_m, and surface tension σ_L:

$$N_{FR} = \frac{v_m^2}{gd} = \frac{(10.06)^2}{(32.2 \times 9)} = 4.186$$
$$\mu_m = 6.27 \times 10^{-4}[\mu_L \lambda + \mu_g(1-\lambda)]$$
$$= 6.27 \times 10^{-4}[2.96(0.72) + 0.0184(0.28)]$$
$$= 1.44 \times 10^{-3} \, lb/(ft./s)$$
$$\sigma_L = 37.5 - 0.257(API) = 37.5 - 0.257(33)$$
$$= 29.0 \, dyn/cm$$

10. Calculate the non-slip Reynolds number and the liquid velocity number:

$$Re_{NS} = \frac{G_m d}{\mu_m} = \frac{(374.4)(9)}{(1.44 \times 10^{-3})12} = 195,000$$
$$N_{LV} = 1.938 \, v_{sL} \left(\frac{\gamma_L}{\sigma_L}\right)^{0.25} = 1.938(7.25)(47.42/29)^{0.25}$$
$$= 15.88$$

11. Calculate $L_1, L_2, L_3,$ and L_4:

$$L_1 = 316\lambda^{0.302} = 316(0.721)^{0.302} = 286$$
$$L_2 = 0.000952\lambda^{-2.4684} = 0.0009252(0.721)^{-2.4684}$$
$$= 0.0021$$
$$L_3 = 0.10\lambda^{-1.4516} = 0.10(0.721)^{-1.4516} = 0.161$$
$$L_4 = 0.5\lambda^{-6.738} = 0.5(0.721)^{-6.738} = 4.53$$

12. Determine flow pattern:

$$\text{Since } 0.721 \geq 0.4 \quad \text{and} \quad L_3 < N_{FR} \leq L_4$$

Flow is intermittent.

13. Calculate the horizontal holdup $H_L(O)$:

$$H_L(O) = a\lambda^b/N_{FR}^c = 0.845 \times 0.721^{0.5351}/7.186^{0.0173}$$
$$= 0.692$$

14. Calculate ψ and $H_L(O)$ and two-phase specific weight:

$$\text{Since } \theta = 0°, \lambda = 1 + 0 = 1$$
$$H_L(0°) = H_L(0)\lambda = 0.692$$
$$\gamma_t = \gamma_L H_L + \gamma_g(1 - H_L) = 47.42(0.692)$$
$$+ 10.96(1 - 0.692)$$
$$= 36.19\,\text{lb/ft.}^3$$

15. Calculate the friction factor ratio:

$$y = \frac{\lambda}{H_L^2} = \frac{0.721}{(0.692)^2} = 1.506$$
$$\ln(y) = 0.4092$$
$$S = \ln(y)/[-0.0523 + 3182\ln y - 0.8725(\ln y)^2$$
$$+ 0.01853(\ln y)^4]$$
$$= 0.3706$$
$$f_t/f_{ns} = e^s = e^{0.3706} = 1.449$$

16. Calculate the non-slip friction factor f_{ns}:

$$f_{ns} = 1/\{2\log[Re_{ns}/(4.5223\log Re_{ns} - 38,215)]\}^2$$
$$= 1/(2\log[195,000/(4.5223\log 195,000 - 3.8215)])^2$$
$$= 0.01573$$

17. Calculate the two-phase friction factor:

$$f_t = f_{ns}(f_t/f_{ns}) = 0.01573(1.449) = 0.0227$$

18. Calculate the pressure gradient:

$$\frac{\Delta L}{\Delta p(144)} = \frac{1 - \dfrac{\gamma_t V_m V_{sg}}{gP}}{\gamma_{tp} \sin\theta + \dfrac{f_t G_m V_m}{2gd}}$$

$$= \frac{1 - \dfrac{(36.19)(10.06)(2.81)}{(32.2)2{,}000(144)}}{36.19(1)(0) + \dfrac{0.0227(374.4)(10.06)}{2(32.2)(9/12)}}$$

$$= 0.5646$$

$$\frac{\Delta L}{P} = 81.3 \quad \text{or} \quad \frac{\Delta p}{\Delta L} = 1.23 \times 10^{-2}\, \text{psi/ft.}$$

because pressure is decreasing in flow direction to proper value of $\Delta p/\Delta L = -1.23 \times 10^{-2}\, \text{psi/ft.}$

The performance of the correlation is given below [7].

- *Tubing Size*: For the range in which the experimental investigation was conducted (i.e., tubing sizes between 1 and 1.5 in.), the pressure losses are accurately estimated. Any further increase in tubing size tends to result in an over prediction in the pressure loss.
- *Oil Gravity*: A reasonably good performance is obtained over a broad spectrum of oil gravities.
- *Gas–Liquid Ratio*: In general, an over predicted pressure drop is obtained with increasing GLR. The errors become especially large for gas–liquid ratio above 5000.
- *Water-Cut*: The accuracy of the pressure profile predictions is generally good up to about 10% water-cut.

2.16 MECHANISTIC MODELS

By the end of the 20th century, mechanistic models had become the focus of many investigations to improve the flow behavior predictions in multiphase flow. These models rely more on the theory or mechanisms (or phenomena) in multiphase flow rather than solely on experimental work. Empiricism is still used in a mechanistic approach to predict certain flow mechanisms or provide closure relationships. One of the preliminary works on mechanistic modeling of flow pattern transition and steady, upward, and gas-liquid flow in vertical tubes is presented by Taitel et al. [16]. This flow pattern prediction was later modified by Barnea et al. [17] to account for a whole range of pipe inclinations. These two works have provided the basis for predicting flow patterns by defining transition boundaries between bubble, slug, churn, and annular flows. Although many mechanistic models predict flow behavior for a single flow pattern, only a few comprehensive mechanistic models such as Hasan and Kabir [18, 19] and Ansari et al. [20] have been published in the literature to model all four flow pattern transitions in multiphase flow. Both these mechanistic the models have enjoyed considerable success in multiphase flow predictions and have generally been accepted in the petroleum industry. The Ansari et al. model [20] is relatively more complex, and the reader is referred to the original papers [20, 21] for a complete presentation of the model. Only the details of the Hasan and Kabir model are presented here.

2.17 HASAN AND KABIR MODEL [18, 19]

Hasan and Kabir developed a mechanistic model to predict multiphase flow pressure gradients and wellbores for the same four flow patterns discussed in the empirical approach.

The bubble/slug transition: The bubble/slug flow transitions across at a void fraction of 0.25 and is expressed in terms of superficial gas velocity.

$$v_{sg} = \frac{\sin\theta}{4 - C_o}(C_o v_{sL} + v_s) \tag{2.73}$$

where C_o is the flow parameter written as

$$C_o = \begin{Bmatrix} 1.2 \text{ if } d < 4 \text{ in} \\ 2.0 \text{ if } d > 4 \text{ in} \end{Bmatrix} \tag{2.74}$$

In equation 2.73, V_s is the bubble-rise or slip velocity and is given by Harmathy [22] expression as

$$v_s = 1.53 \left[\frac{g\sigma_L(\rho_L - \rho_g)}{\rho_L^2} \right]^{1/4} \tag{2.75}$$

The transition to slug flow occurs at superficial gas velocities greater than that given by equation 2.73. The slug flow transition is also checked by finding the rise velocity of a Taylor bubble as

$$v_{TB} = 0.35 \sqrt{\frac{g\sigma_L(\rho_L - \rho_g)}{\rho_L^2}} \sqrt{\sin\theta}(1 + \cos\theta)^{1.2} \tag{2.76}$$

It should be noted that v_{TB} is dependent on the pipe diameter. In smaller diameter pipes, when Taylor bubble velocity is less than in the slip velocity, the rising smaller bubbles approached the back of the Taylor bubble, coalesce with it, and increase its size thus causing a transition to slug flow.

Dispersed bubble/slug/churn flow transitions: Taitel et al. [16] proposed an equation for the transition to dispersed-bubble flow and is given by

$$v_m^{1.12} = 4.68 d^{0.48} \sqrt{\frac{g\sigma_L(\rho_L - \rho_g)}{\rho_L^2}} \left(\frac{\sigma_L}{\rho_L}\right)^{0.6} \left(\frac{\rho_L}{\mu_L}\right)^{0.08} \quad (2.77)$$

However, as the gas void fraction exceeds 0.52, bubble coalescence cannot be prevented, and transition to either slug, churn, or annular flow must occur.

Slug/churn transitions is predicted from

$$\rho_g V_{sg}^2 = 0.0067(\rho_L V_{sL}^2)^{1.7} \text{ if } \rho_L V_{sL}^2 < 50 \quad (2.78a)$$

and $\quad \rho_g V_{sg}^2 = 17.1 \log_{10}(\rho_L V_{sL}^2) - 23.2 \text{ if } \rho_L V_{sL}^2 \geq 50 \quad (2.78b)$

Annular-flow transition: The transition to annular flow is given by

$$V_{sg} = 3.1 \left[\frac{g\sigma_L(\rho_L - \rho_g)}{\rho_g^2}\right]^{1/4} \quad (2.79)$$

This criterion is partly based on the gas velocity required to prevent fallback of entrained liquid droplets in the gas stream.

Bubble and dispersed bubble flow: The liquid hold up for the bubble and dispersed bubble flow is given by

$$H_L = 1 - \frac{V_{sg}}{C_o V_m + V_s} \quad (2.80)$$

The total pressure gradient is estimated by Equation 2.81, and it is used in conjunction with the mixture density, ρ_m, once the hold up is calculated.

$$\left(\frac{dp}{dL}\right)_t = \rho_m g \sin\theta + \frac{2f\rho_m v_m^2}{d} + \rho_m v_m \frac{dv_m}{dL} \quad (2.81)$$

where $\rho_m = H_L \rho_L + (1 - H_L)\rho_g$

As usual, the friction factor is determined from the Moody diagram shown in Figure 2.32 [11]. Mixture Reynolds number for determining the friction factor is given by the Equation 2.82.

$$Re_m = \frac{\rho_L v_m d}{\mu_L} \quad (2.82)$$

Slug and churn flow: The slip velocity for expressions for slug and churn flow is given by Equation 2.83

$$V_s = 0.35 \sqrt{\frac{gd(\rho_L - \rho_g)}{\rho_L}} \sqrt{\sin\theta}(1 + \cos\theta)^{1.2} \quad (2.83)$$

A value of 1.2 and 1.15 are used for C_o in Equation 2.80 for slug and churn flow, respectively. The friction pressure gradient is calculated from Equation 2.81. Because of the chaotic nature of the flow and difficulty in analyzing the churn flow regime, predictions in this flow regime are less accurate.

Annular flow: The liquid hold up for the central core in annular flow is given by

$$H_{LC} = \frac{Ev_{sL}}{Ev_{sL} + v_{sg}} \qquad (2.84)$$

Here $H_{LC} = 1 - f_{gc}$ where f_{gc} is the gas void fraction and E is the liquid entrainment in the central core. The liquid entrainment in the core is calculated by the following expressions

$$\left\{ \begin{array}{l} E = 0.0055[(v_{sg})_{crit} \times 10^4]^{2.86} \text{ if } (v_{sg})_{crit} \times 10^4 < 4 \\ E = 0.857 \log_{10}[(v_{sg})_{crit} \times 10^4] - 0.20 \text{ if } (v_{sg})_{crit} \times 10^4 > 4 \end{array} \right\} \qquad (2.85)$$

$(v_{sg})_{crit}$ is given by

$$(v_{sg})_{crit} = \frac{v_{sg}\mu_g}{\sigma_L} \left(\frac{\rho_g}{\rho_L} \right)^{1/2} \qquad (2.86)$$

The friction pressure gradient is calculated from

$$\left(\frac{dp}{dL} \right)_f = \frac{2f_C\rho_C}{d} \left(\frac{v_{sg}}{f_{gc}} \right)^2 \qquad (2.87)$$

where

$$f_C = \frac{0.079(1 + 75H_{LC})}{N_{Reg}^{0.25}} \qquad (2.88)$$

and

$$\rho_C = \frac{v_{sg}\rho_g + v_{sL}\rho_L E}{v_{sg} + v_{sL}E} \qquad (2.89)$$

Wallis [3] presented a simple equation for gas void fraction in the core given by

$$f_{gc} = (1 + X^{0.8})^{-0.378} \qquad (2.90)$$

where X is the Lockhart-Martinelli parameter given in terms of the gas mass fraction x_g as

$$X = \left[\frac{(1 - x_g)}{x_g} \right]^{0.9} \left(\frac{\mu_L}{\mu_g} \right)^{0.1} \sqrt{\frac{\rho_L}{\rho_g}} \qquad (2.91)$$

Example 2.7
Solve Example 2.2 using the Hasan-Kabir mechanistic model.

The bubble/slug transition is checked using Equation 2.73. The slip velocity is first found Equation 2.75

$$v_s = 1.53 \left[\frac{g\sigma_L(\rho_L - \rho_g)}{\rho_L^2} \right]^{1/4}$$

$$v_s = 1.53 \left[\frac{32.17^2(0.001918)(47.6 - 8.823)}{47.6^2} \right]^{1/4}$$

$$= 0.656 \, ft/s$$

Using $C_o = 1.2$, V_{sg} is calculated from Equation 2.73 as

$$V_{sg} = \frac{\sin\theta}{4 - C_o}(C_o v_{sL} + v_s)$$

$$V_{sg} = \frac{1}{4 - 1.2}(1.2 \times 2.147 + 0.656)$$

$$= 1.1544 \, ft/s$$

Since $V_{sg} > 0.656$ ft/s, bubble flow does not exist. Now, check dispersed-bubble transition (Equation 2.77)

$$v_m^{1.12} = 4.68 \left(\frac{d}{12} \right)^{0.48} \sqrt{\frac{g(\rho_L - \rho_g)}{g_c\sigma_L}} \left(\frac{g_c\sigma_L}{\rho_L} \right)^{0.6} \left(\frac{\rho_L}{g_c\mu_L} \right)^{0.08}$$

$$\left(\frac{dP}{dL} \right)_f = \frac{2f\rho_L v_m^2 H_L}{d}$$

$$= \frac{2 \times 0.004 \times 47.6 \times 6.227^2 \times 0.5025}{0.1662}$$

$$= 1.38645 \, lb_f/ft^3 = 0.009628 \, psi/ft$$

Thus, the total pressure gradient

$$\frac{dP}{dL} = 0.1965 + 0.009628 = 0.20618 \, psi/ft$$

$$\Delta L = \frac{500}{0.20618} = 2425.07 \, ft$$

In general, the performance of the Hasan-Kabir mechanistic model is good for a wide range of tubing sizes, oil gravity, gas–liquid ratio, and water-cut.

2.18 SUMMARY

In this work, attention was paid only to six methods: flow regime maps, the Duns-Ros, Orkiszewski, Hagedorn-Brown, Beggs-Brill, and Hasan-Kabir methods. They are the most often used. A comprehensive discussion of the many other multiphase models published in the literature, including the ones discussed here, can be found in a published monograph [23]. As alluded to in the preceding discussions, the pressure profile in a pipe is influenced by a number of parameters such as tubing size and geometry, well inclination, and fluid composition and properties (which in turn depend on pressure and temperature). Therefore, the empirical methods presented above should be used with caution, keeping in mind the experimental range of validity for which the correlation have been developed. For example, the Hagedorn-Brown, Orkiszewski, and Duns-Ros correlations have all been developed for vertical wells, and their use in deviated wells may or may not yield accurate results. In addition, the Duns-Ros correlation is not applicable for wells with water-cut and should be avoided for such cases. Similarly, the Beggs-Brill method, primarily developed for inclined wells, is applicable to multiphase flows with or without water cut. In general, mechanistic models such as the Hasan-Kabir have a wide range of applicability because of their limited reliance on empiricism.

Over the years, numerous modifications to the above discussed empirical models have been proposed to improve their multiphase flow predictions and to extend their range of applicability. Some of these modifications are discussed in the monograph by Brill and Mukherjee [23] and are used extensively in many commercial software programs. The monograph also presents a comparative study of the performance of both empirical and mechanistic models with measured field data from a well databank for a variety of cases. Although some methods/models perform better than others for certain cases, no single method performs accurately for all cases despite development of new mechanistic models. Therefore, it is strongly recommended that the model limitations and validity be kept in mind before a particular model/correlation is selected in the absence of other relevant information such as measured field data.

References

[1] Brown, K. E., *The Technology of Artificial Lift Methods*, Vol. 1, Pennwell Books, Tulsa, Oklahoma, 1977.

[2] Golan, M., and Whitson, C. H., *Well Performance*, Prentice Hall, Englewood Cliffs, New Jersey, 1991.

[3] Wallis, G. B., *One-Dimensional Two-Phase Flow*, McGraw-Hill Book Co., New York, 1969.

[4] Crowley, Ch. J., Wallis, G. B., and Rothe, P. H., *State of the Art Report on Multiphase Methods for Gas and Oil Pipelines*, Vols. 1–3, AGA Project, PR-172-609, December 1986.

[5] Duns, Jr. H., and Ros, N. C. J., "Vertical Flow of Gas and Liquid Mixtures in Wells," *Proc. Sixth World Pet. Congress*, Frankfurt, Section II, 22-PD6, June 1963.

[6] Lawson, J. D., and Brill, J. P., "A Statistical Evaluation of Methods Used to Predict Pressure Losses for Multiphase flow in Vertical Oilwell Tubing," *J. Pet. Tech.*, August 1974.

[7] Vohra, I. R., Robinson, J. R., and Brill, J. P., "Evaluation of Three New Methods for Predicting Pressure Losses in Vertical Oilwell Tubing," *J. Pet. Tech.*, August 1974.

[8] Orkiszewski, J., "Predicting Two-Phase Pressure Drops in Vertical Pipe," *J. Pet. Tech.*, June 1967.

[9] Byod, O. W., *Petroleum Fluid Flow Systems*, Campbell Petroleum Series, Norman, Oklahoma, 1983.
[10] Griffith, P. G., and Wallis, G. B., "Two-Phase Slug Flow," *ASME Journal of Heat Transfer*, 1967.
[11] Moody, L. F., "Friction Factors in Pipe Flow," *Transactions of ASME*, November 1941.
[12] Hagedorn, A. R., and Brown, K. E., "The Effect of Liquid Viscosity in Vertical Two-Phase Flow," *Journal of Petroleum Technology*, February 1964.
[13] Hagedorn, A. R., and Brown, K. E., "Experimental Study of Pressure Gradients Occurring During Continuous Two-Phase Flow in Small Diameter Vertical Conduits," *J. Pet. Tech.*, April 1965.
[14] Barker, A., Nielson, K., and Galb, A., "Pressure Loss Liquid-Holdup Calculations Developed," *Oil & Gas Journal*, March 1988 (three parts).
[15] Beggs, H. D. and Brill, J. P., "A Study of Two-Phase Flow in Inclined Pipes," *J. Pet. Tech.*, May 1973.
[16] Taitel, Y. M., Barnea, D., and Dukler, A. E., "Modeling Flow Pattern Transitions for Steady Upward Gas-Liquid Flow in Vertical Tubes," *AIChE J*. 26, 1980.
[17] Barnea, D., Shoham, O., and Taitel, Y., "Flow Pattern Transition for Vertical Downward Two-Phase Flow," *Chem, Eng. Sci.* 37, 1982.
[18] Hasan, A. R., and Kabir, C. S., "A Study of Multiphase Flow Behavior in Vertical Wells," *SPEPE 263 Trans.*, AIME, May 1998.
[19] Hasan, A. R., and Kabir, C. S., "Predicting Multiphase Flow Behavior in a Deviated Well," *SPEPE 474*, November 1988.
[20] Ansari, A. M., Sylvester, N. D., Shoham, O., and Brill, J. P., "A Comprehensive Mechanistic Model for Two-Phase Flow in Wellbores," *SPEPF Trans.*, AIME, May 1994.
[21] Ansari, A. M., Sylvester, N. D., Sarica, C., et al. "Supplement to paper SPE 20630, A Comprehensive Mechanistic Model for Upward Two-Phase Flow in Wellbores," paper SPE 28671, May 1994.
[22] Harmathy, T. Z., "Velocity of Large Drops and Bubbles in Media of Infinite or Restricted Extent," *AIChE J.* 6, 1960.
[23] Brill, J. P., and Mukherjee, H.: "Multiphase Flow in Wells," SPE Monograph Volume 17, Henry L. Doherty Series, Richardson, Texas, 1999.

3

Natural Flow Performance

The most important parameters that are used to evaluate performance or behavior of petroleum fluids flowing from an upstream point (in reservoir) to a downstream point (at surface) are pressure and flow rate. According to basic fluid flow through reservoir, production rate is a function of flowing pressure at the bottomhole of the well for a specified reservoir pressure and the fluid and reservoir properties. The flowing bottomhole pressure required to lift the fluids up to the surface may be influenced by the size of the tubing string, choke installed at downhole or surface and pressure loss along the pipeline.

In oil and gas fields, the flowing system may be divided into at least four components:

1. reservoir
2. wellbore
3. chokes and valves
4. surface flowline

Each individual component, through which reservoir fluid flow, has its own performance and, of course, affects each other. A good understanding of the flow performances is very important in production engineering. The combined performances are often used as a tool for optimizing well production and sizing equipment. Furthermore, engineering and economic judgments can depend on good information on the well and reasonable prediction of the future performances.

As has been discussed in previous chapters, hydrocarbon fluids produced can be either single phase (oil or gas) or two phases. Natural flow performance of oil, gas, and the mixture will therefore be discussed separately. Some illustrative examples are given at the end of each subsection.

3.1 INFLOW PERFORMANCE

Inflow performance represents behavior of a reservoir in producing the oil through the well. For a heterogeneous reservoir, the inflow performance might differ from one well to another. The performance is commonly defined in terms of a plot of surface production rate (stb/d) versus flowing bottomhole pressure (P_{wf} in psi) on cartesian coordinate. This plot is defined as inflow performance relationship (IPR) curve and is very useful in estimating well capacity, designing tubing string, and scheduling an artificial lift method.

For single-phase liquid flow, radial flow equation can be written as (for oil)

1. semi–steady-state condition

$$q_0 = 0.00708 k_0 h \frac{(P_r - P_{wf})}{\overline{\mu}_0 \overline{B}_0 \left(\ln \dfrac{r_e}{r_w} - \dfrac{3}{4} + s \right)} \tag{3.1a}$$

2. steady-state condition

$$q_0 = 0.00708 k_0 h \frac{(P_r - P_{wf})}{\overline{\mu}_{\overline{B}_0} \left(\ln \dfrac{r_e}{r_w} - \dfrac{1}{2} + s \right)} \tag{3.1b}$$

where q_0 = surface measured oil rate in stb/d
   k_0 = permeability to oil in md
   h = effective formation thickness in ft.
   P_r = average reservoir pressure in psia
   P_{wf} = flowing bottomhole pressure in psia
   $\overline{\mu}_0$ = oil viscosity evaluated at $\dfrac{(P_r + P_{wf})}{2}$ in cp
   $\overline{B}_0$ = oil formation volume factor evaluated at
       $\dfrac{(P_r + P_{wf})}{2}$ in bbl/stb
   r_e = drainage radius in ft.
   r_w = wellbore radius in ft.
   s = skin factor, dimensionless

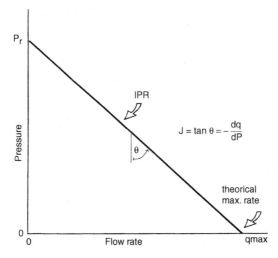

FIGURE 3.1 Inflow performance relationship of single-phase oil reservoirs.

Assuming all parameters but P_{wf} are constants in the equations above, it is also clear the flowrate q_0 is linearly proportional to flowing pressure P_{wf}. Therefore, for laminar flow the plot q_0 versus P_{wf} on a cartesian coordinate must be linear. This is illustrated in Figure 3.1. Strictly speaking, it shows the behavior of single-phase liquid flowing over the range of P_{wf}. In actual cases, however, straightline IPR may be shown by reservoirs producing at P_r and P_{wf} above the bubble point pressure P_b, and by strong water-drive reservoirs.

Productivity index, usually denoted by the symbol J, is commonly expressed in practice for well performance. It is mathematically defined as

$$J = \frac{q_0}{P_r - P_{wf}} \tag{3.2}$$

where J is in stb/d/psi. The term $(P_r - P_{wf})$ is called *pressure drawdown*. Equation 3.1a or 3.1b can be rearranged to be used in estimating well productivity index.

By knowing reservoir pressure P_r, it is possible to construct an oil IPR curve from a single flow test on a well. Or, because of the linearity of liquid IPR curves, by conducting a two-point flow test (two different flowrates while measuring the flowing bottomhole pressure) on a well, the static reservoir pressure can be determined.

The equations discussed above are derived from the laminar Darcy's law. In a case where turbulent flow occurs, a modified equation should be used. The occurrence of turbulence at the bottomhole may indicate too few open perforations or too narrow fracture in fractured well or other incorrect completion method applied. All these bring about inefficient production operation because the high drawdown encountered results in insufficient flowrate. The symptom may be analyzed using the correlation of Jones et al. [1]:

$$\frac{\Delta P}{q_0} = C + Dq_0 \tag{3.3}$$

where

$$C = \frac{\mu_0 B_0}{0.00708 k_0 h} \left(\ln \frac{r_e}{r_w} - \frac{3}{4} + s \right) \tag{3.4}$$

is called the laminar flow coefficient, and

$$D = \frac{9.08 \times 10^{-13} \beta B_0^2 \gamma_0}{4\pi^2 h^2 r_w} \tag{3.5}$$

is the turbulence coefficient, with β = the turbulence factor in ft.$^{-1}$ and γ_0 = oil specific weight in lb/ft.^{3} and other terms are the same as in the previous equations. The magnitude of the turbulent factor is in the order of 10^{-6} and 10^{-8} and is usually negligible when compared with the laminar flow coefficient in most oil wells. But if this is not the case, plot ($\Delta P/q$) versus q on a cartesian coordinate paper. If the flow is fully laminar, then the plot has a slope of zero. But when turbulence is measurable, the plot has non-zero positive slope, which also means that the productivity decreases as flowrate increases.

3.1.1 Predicting Future Oil Well IPR

Pertaining to our problem here dealing with single-phase oil flow in reservoirs, we always assume that gas does not develop over the whole range of flowing pressure at downhole. The consequence is that the following equations are valid for wells that produce only oil (and water).

Recalling the radial flow equation for oil (Equation 3.1a,b for instance), we obtain

$$q_{0max} = \frac{0.00708 k_0 h P_r}{\bar{\mu}_0 \bar{B}_0 \left(\ln \frac{r_e}{r_w} - \frac{3}{4} + s \right)} \tag{3.6}$$

where q_{0max} is a theoretical possible maximum flowrate when $P_{wf} \approx 0$.

Assuming no changes in producing interval, skin factor and drainage radius occur during a period of time from present to the future, and also $\bar{\mu}_0$ and $\bar{B}_0$ are nearly constant over the whole range of pressure, the future possible maximum flowrate is:

$$(q_{0max})_f = (q_{0max})_p \times \frac{(k_{r0} \times P_r)_f}{(k_{r0} \times P_r)_p} \tag{3.7}$$

Because no gas develops in the reservoir, the relative permeability to oil can be a function of water saturation. Figure 3.2 suggests the possibility of changes in oil inflow performance curves with time ($t_2 > t_1$).

3.1.2 Tubing Performance

A tubing performance may be defined as the behavior of a well in giving up the reservoir fluids to the surface. The performance is commonly showed as a plot of flowrate versus flowing pressure. This plot is called the tubing performance relationship (TPR). For a specified wellhead pressure, the TPR curves vary with diameter of the tubing. Also, for a given tubing size, the curves vary with wellhead pressure. Figure 3.3 shows the effect of tubing size and wellhead pressure [2].

For single-phase liquid flow, pressure loss in tubing can be determined using a simple fluid flow equation for vertical pipe, or using some graphical pressure loss correlations where available with GLR = 0.

Tubing performance curves are used to determine the producing capacity of a well. By plotting IPR and TPR on the same graph paper, a stabilized maximum production rate of the well can be estimated. Figure 3.3 shows the combined plots for determining the flowrate. The larger the diameter of tubing, the higher the flowrate that can be obtained. But there is a critical diameter limiting the rate, even lowering the well capacity. For a specified tubing size, the lower the wellhead pressure, the higher the production rate.

3.1.3 Choke Performance

A choke can be installed at the wellhead or downhole to control natural flow or pressure. Chokes are widely used in oil fields. Several reasons in installing chokes are to regulate production rate, to protect surface equipments from slugging, to avoid sand problem due too high drawdown, or to control flowrate in order to avoid water or gas coning.

There are two types of wellhead choke that are commonly used, positive chokes and adjustable chokes. A positive choke has a fixed size in diameter so that it must be replaced to regulate production rate. An adjustable choke permits gradual changes in the size of the opening.

Placing a choke at the wellhead can mean fixing the wellhead pressure and thus flowing bottomhole pressure and production rate. For a given wellhead pressure, by calculating pressure loss in the tubing, the flowing bottomhole pressure can be determined. If reservoir pressure and productivity index of the well are known, the flowrate can then be determined using Equation 3.2.

The rate of oil flowing through a choke (orifice or nozzle) depends upon pressure drop in the choke, the inside diameters of pipe and choke, and

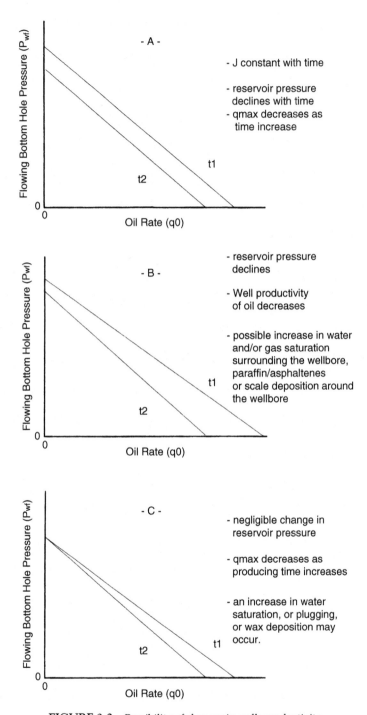

FIGURE 3.2 Possibility of changes in wells productivity.

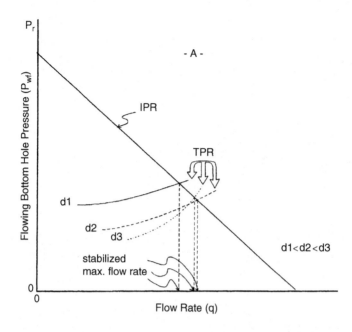

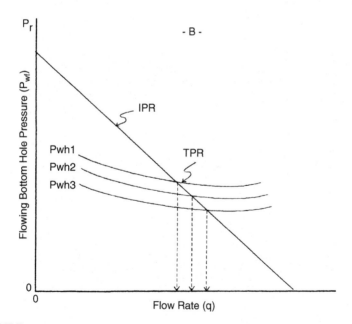

FIGURE 3.3 Effects of tubing size and tubing head pressure on a well productivity.

density of the oil. For incompressible fluids, the Equation 3.8 may be used to estimate the flowrate of oil:

$$q_0 = 10,285CA\sqrt{\frac{\Delta P}{\gamma_0}} \qquad (3.8)$$

where q_0 = oil rate in bbl/day
$\quad$ C = flow coefficient as function of diameter ratio and Reynolds number (see Figure 3.4)
$\quad$ A = cross-sectional area of choke in in.^{2}
$\quad$ ΔP = pressure drop across the choke in psi
$\quad$ γ_0 = oil specific weight in lb/ft.^{3}

In installing a choke, the downstream pressure of the choke is usually 0.55 of the upstream pressure, or even less to ensure no change in flowrate or upstream pressure. This condition is called a sonic flow. A subsonic flow occurs when the upstream pressure or flowrate is affected by a change in downstream pressure.

3.1.4 Flowline Performance

After passing through a choke installed at the wellhead, the oil flows through flowline to a separator. If the separator is far from the wellhead and the pressure loss in the flowline cannot be neglected, pressure-flowrate relationship for flowline can be generated similar to tubing performance curves. Usually the separator pressure is specified. Then by using pressure gradient curves available for horizontal pipes or using a simple horizontal fluid flow equation, the wellhead pressure or downstream pressure of the choke or intake pressure of the flowline can be determined as a function of flowrate. This pressure-flowrate plot is useful in sizing the flowline. Figure 3.5 illustrates the relationship between the wellhead pressure and flowrate for some different flowline diameters. This plot is called *flowline performance curve*.

By plotting TPR in term of wellhead pressure for various tubing sizes and flowline performance curves on the same graph (Figure 3.6), selecting tubing string-flowline combination for a well can be established based on the pipe's availability, production scheme planned, and economic consideration.

Example 3.1
Determination of oil inflow performance.

Suppose two flowrates are conducted on an oil well. The results are as follows:

	Test 1	Test 2
q_0, stb/d	200	400
P_{wf}, psi	2400	1800

Gas/oil ratios are very small. Estimate the reservoir pressure and productivity index of the well, and also determine the maximum flowrate.

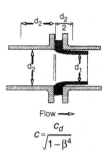

$$C = \frac{C_d}{\sqrt{1 - \beta^4}}$$

Example: The flow coefficient C for a diameter ratio β of 0.60 at a Reynolds number of 20,000 (2×10^4) equal 1.03

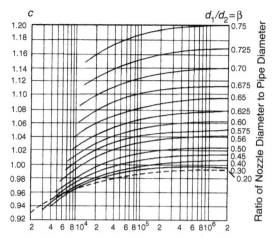

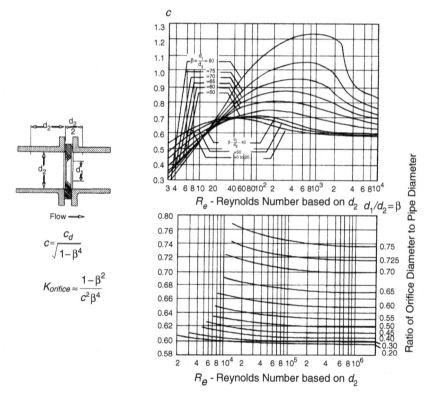

$$C = \frac{C_d}{\sqrt{1 - \beta^4}}$$

$$K_{orifice} \approx \frac{1 - \beta^2}{c^2 \beta^4}$$

FIGURE 3.4 Flow coefficient versus diameter ratio and Reynolds number [2].

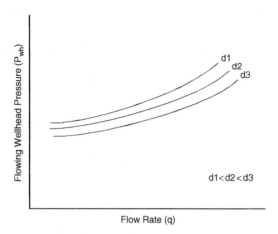

FIGURE 3.5   Flowline performance curves for different flowline diameters.

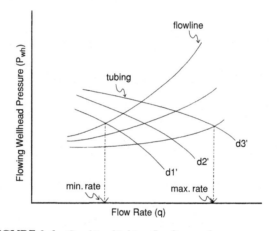

FIGURE 3.6   Combined tubing-flowline performance curves.

Solution

1. Plot the two data points on a Cartesian graph paper (q_0 versus P_{wf}), see Figure 3.7. Draw a straight line through these two points, the intersection with ordinate is the estimated reservoir pressure which is about 3000 psi.

2. The productivity index is

$$J = \frac{200}{3,000 - 2,400} = 0.333 \text{ stb/d/psi}$$

3. The theoretical maximum flowrate is:

$$q_{max} = 0.333(3,000 - 0) = 1,000 \text{ stb/d}$$

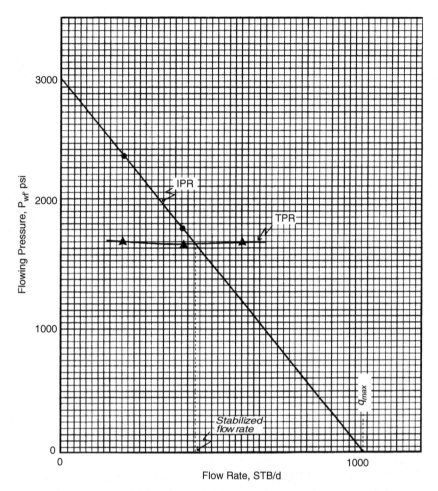

FIGURE 3.7 Pressure-rate relationship for examples A and B on pages 542 to 544.

Example 3.2

The well illustrated in Example 3.1 has a vertical depth of 4100 ft. Tubing string of $2\frac{3}{8}$ in. has been installed. The flowing pressure at wellhead is 210 psi. What is the stabilized oil rate achieved?

Solution

Using Gilbert's correlation, we obtain:

q, stb/d	P_{wf}, psi
200	1700
400	1695
600	1690

Plot these values of q and P_{wf} on the same graph paper used for solving problem Example 3.1 above (Figure 3.7). What we get is the TPR curve intersecting the IPR curve. This intersection represents the stabilized rate achieved, which is 435 stb/d.

Example 3.3

The oil well of Example 3.1 is producing 25° API oil. If a positive choke with diameter of 14/64 in. is installed at the wellhead, determine the pressure at the downstream of the choke.

Solution

$$\gamma_0 = \frac{141.5}{131.5+25} \times 62.4 = 56.42 \text{ lb/ft.}^3$$

Rearranging Equation 3.8 and assuming that $C \approx 1.0$

$$\Delta P = 56.42 \times \left(\frac{435}{10,285(1.0)(0.03758)} \right)^2 = 72 \text{ psi}$$

$$P_{downstream} = (210 - 72) \text{ psi} = 138 \text{ psi}$$

3.1.5 Gas Flow Performances

As for oil wells, performance curves characterizing a gas production system are very useful tools used to visualize and graphically predict the effects of declining reservoir pressure, changes in tubular size, increasing water production, or installing gas compressors.

3.1.6 Gas Inflow Performance

A mathematical expression commonly used to relate gas flowrate and flowing bottomhole pressure is

$$q = C(P_r^2 - P_{wf}^2)^n \tag{3.9}$$

where q = gas flowrate in Mscf/d
$\quad$ P_r = shut-in reservoir pressure in psia
$\quad$ P_{wf} = flowing bottomhole pressure in psia
$\quad$ C = stabilized performance coefficient, constant
$\quad$ n = numerical exponent, constant

Equation 3.9 was first introduced by Rawlins and Schellhardt [3] in 1935 and is known as the back-pressure equation. From gas well test data, plotting q versus $(P_r^2 - P_{wf}^2)$ on a log-log graph will give a straight line passing through the data points, see Figure 3.8. This plot was made based on a stabilized four-point test.

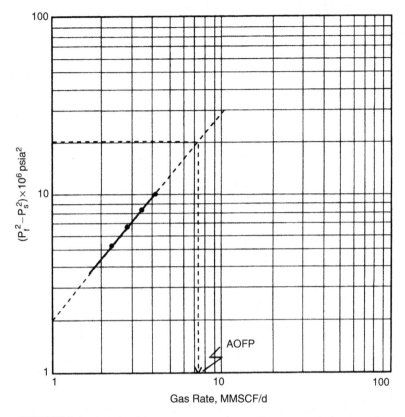

FIGURE 3.8 Stabilized four-point test and open flow potential of a gas well.

The information that can be obtained from this plot is the absolute open flow potential (AOFP) of the well. This is defined as the theoretical maximum flowrate when flowing pressure at the sand face is zero.

Determination of the exponent n and the coefficient C is given here using Figure 3.8 as follows:

1. Choose two values of q arbitrarily but separated one cycle from each other.
2. Read corresponding values of $(P_r^2 - P_{wf}^2)$.
3. Calculate

$$n = 1/\text{slope} = \frac{\log q_2 - \log q_1}{\log(P_r^2 - P_{wf}^2)_2 - \log(P_r^2 - P_{wf}^2)_1}$$

Choosing $q_1 = 1$ gives $(P_r^2 - P_{wf}^2)_1 = 1.5$, and $q_2 = 10$ gives $(P_r^2 - P_{wf}^2)_2 = 20.7$

Then

$$n = \frac{\log 10 - \log 1}{\log 20.7 - \log 1.5} = 0.877$$

4. Rearranging Equation 3.9, we obtain:

$$C = \frac{q}{(P_r^2 - P_{wf}^2)^n} = \frac{10,000}{(20.7 \times 10^6)^{0.877}}$$
$$= 3.84 \times 10^{-3} \text{ Mscf/d/psia}^{2n}$$

The AOFP of the well can then be calculated as:

$$AOFP = q_{max} = 3.84 \times 10^{-3}(3887^2 - 0^2)0.877$$
$$= 7585 \text{ Mscf/d}$$
$$= 7.585 \text{ MMscf/d}$$

From the graph, the AOFP $= 7.6$ MMscf/d

The IPR curve can be constructed by using the deliverability equation above. By taking some values of P_{wf} arbitrarily, the corresponding q's can be calculated. The IPR curve for the example is shown in Figure 3.9.

For situations where multipoint tests cannot be run due to economic or other reasons, single-point data can be used to generate the IPR curve provided that a shut-in bottomhole pressure is known. Mishra and Caudle [4] proposed a simple method for generating a gas IPR curve from just a single-point test data. Employing the basic gas flow in term of pseudo-pressure function, they developed a dimensionless IPR curve to be used as a reference curve. As an alternative, the dimensionless IPR equation to the best-fit curve is introduced

$$\frac{q}{q_{max}} = \frac{5}{4}\left(1 - 5^{\frac{m(P_{wf})}{m(P_r)} - 1}\right) \tag{3.10}$$

where q_{max} = AOFP in Mscf/d
 $m(P_{wf})$ = pseudo-pressure function for real gas and defined as

$$m(P_{wf}) = 2\int_{P_b}^{P_{wf}} \left(\frac{P}{\mu Z}\right) dP$$

$$m(P_r) = 2\int_{P_b}^{P_r} \left(\frac{P}{\mu Z}\right) dP \tag{3.11}$$

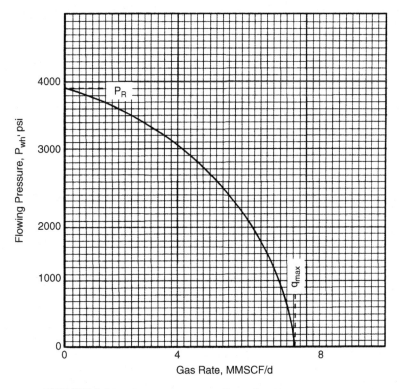

FIGURE 3.9 Inflow performance relationship for a gas reservoir.

where μ = gas viscosity(function of P at isothermal condition) in cp

Z = gas compressibility factor, dimensionless

The use of pseudo-pressure function is quite complex. A numerical integration technique, however, can be applied to this problem. A detail example in applying this numerical technique can be found in *Fundamentals of Reservoir Engineering* [5].

Later, Chase and Anthony [6] offered a simpler method that is a modification to the Mishra-Caudle method. The method proposed involves substitution of real pressure P or P^2 for the real gas pseudopressure function m(P). The squared pressure P^2 is used for pressures less than approximately 2100 psia, and the relevant equation is

$$\frac{q}{q_{max}} = \frac{5}{4}\left(1 - 5^{\left(\frac{P_{wf}^2}{P_r^2} - 1\right)}\right) \tag{3.12}$$

The real pressures P is suggested for pressures greater than approximately 2900 psia. By having the average reservoir pressure P_r, and a single-point test data P_{wf} and q, it is possible to determine the AOFP and to generate the inflow performance curve:

$$\frac{q}{q_{max}} = \frac{5}{4}\left(1 - 5^{\left(\frac{P_{wf}}{P_r} - 1\right)}\right)$$
(3.13)

For pressures ranging for 2100 to 2900 psia, the original Mishra-Caudle's technique is recommended.

3.1.7 Low-Permeability Well Tests

The requirement of the back-pressure method of testing is that the data be obtained under stabilized conditions. That means that the coefficient C of Equation 3.9 is constant with time. This coefficient depends on reservoir characteristics, extent of drainage radius, and produced fluid characteristics.

Wells completed in highly permeable formations stabilize quickly. As demand for gas increased over the years, wells were completed in less permeable formations. In wells of this type, the stabilization period may be very long. Therefore, methods were needed that would permit testing of this type of well without undue waste of time.

In 1955 M. H. Cullender described the isochronal method for determining flow characteristics [7]. The method is based on the assumption that the slope of performance curves of gas wells, exponent n of Equation 3.9, is independent of the drainage area. It is established almost immediately after the well is opened. However, the performance coefficient C decreases with time as the radius of drainage recedes from the well. When the radius reaches the boundary of the reservoir or the area of interference of another well, C becomes a constant and the flow is stabilized.

Under the method the well is opened, the flow and pressure data are obtained at specific time intervals without changing the rate of flow. The well is then closed in until the shut-in pressure is reached, approximately the same as at the beginning of the first test. The well is then opened and produced at a different rate, and the pressure and flow data are collected. This procedure is repeated as many times as desired.

Plotting of these data on log-log paper results in a series of parallel lines, the slope of which gives the coefficient n. This is illustrated in Figure 3.10. Relationship of coefficient C and time for a gas well is illustrated in Figure 3.11.

From these test and theoretical considerations, different procedures have been developed that permit prediction of the coefficients of performance of gas wells produced from low-permeability formations.

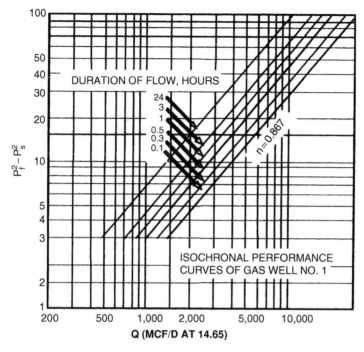

FIGURE 3.10 Isochronal performance curves of gas well no. 1.

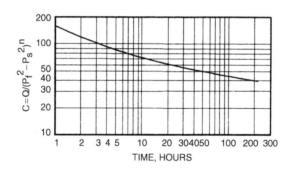

FIGURE 3.11 Relationship of coefficient of performance and time of gas well no. 3.

3.1.8 Predicting Future IPR

Predicting the wells' deliverability is important to be able to plan some changes required to maintain the production capacity. Here simple but reliable methods are introduced to forecasting future inflow performance gas wells.

Accompanying Equation 3.10, Mishra and Caudle presented an empirical equation for predicting gas wells productivity [4]. The equation is:

$$\frac{(q_{max})_f}{(q_{max})_p} = \frac{5}{3}\left(1 - 0.4^{\frac{m(P_r)_f}{m(P_r)_p}}\right) \tag{3.14}$$

where subscripts f and p refer to future and present time, respectively. Later, Chase and Anthony [6] also proposed the simplified form of Equation 3.14, by substituting real pressure for pseudo-pressure function.

To estimate the AOFP of a gas well, one does not have to run well tests as discussed above. An alternative method is to calculate bottomhole pressure, static and flowing, without running a pressure gage down into the well, by knowing pressures at the wellhead. The calculation of flowing bottomhole pressures will be discussed later. Below, the equations for calculating static bottomhole pressure are given.

One of the most common methods in estimating static bottomhole pressure of a gas well is that of Cullender and Smith, which treats the gas compressibility factor as a function of depth [8]. If we divide the well by equal length, one can calculate the static bottomhole pressure as follows:

1. Having a knowledge of static wellhead pressure, pressure at midpoint of the well is calculated by trial and error:

$$0.0375G_g\frac{H}{2} = (P_{ms} - P_{ws})(I_{ms} - I_{ws}) \tag{3.15}$$

where G_g = gas gravity (air = 1.0)
       H = well depth in ft.

$$I_{ms} = (T_{ms} \times Z_{ms})/P_{ms} \tag{3.16}$$

where T_{ms} = absolute temperature at midpoint in °R
       P_{ms} = pressure at midpoint (assumed to calculate Z_{ms}) in psia
       Z_{ms} = gas compressibility factor evaluated at T_{ms} and P_{ms}
       P_{ws} = static wellhead pressure in psia

$$I_{ws} = \frac{T_{ws}Z_{ws}}{P_{ws}} \tag{3.17}$$

where T_{ws} = absolute temperature at wellhead in °R
       Z_{ws} = gas compressibility factor evaluated at T_{ws} and P_{ws}

The problem here is to calculate P_{ms}. If calculated $P_{ms} \approx P_{ms}$ assumed to determine Z_{ms}, then calculation of bottomhole pressure is the next step. If not, use calculated P_{ms} to determine new Z_{ms} and again use Equations 3.16 and 3.17 to calculate a new P_{ms}. Repeat this procedure until calculated P_{ms} is close to assumed P_{ms}.

2. The same procedure is used, and the equation for static bottomhole pressure P_{bs} is:

$$0.0375G_g\frac{H}{2} = (P_{bs} - P_{ms})(I_{bs} + I_{ms}) \tag{3.18}$$

3. The very last step is to apply Simpson's rule to calculate P_{bs}:

$$0.0375G_gH = \frac{P_{bs} - P_{ws}}{3}(I_{ws} + 4I_{ms} + I_{bs}) \tag{3.19}$$

For wells producing some liquids, the gas gravity G_g in Equations 3.15 and 3.18 must be replaced by γ_{mix}:

$$G_{mix} = \frac{G_g + 4584G_0/R}{1 + 132800G_0/RM_0} \tag{3.20}$$

where G_{mix} = specific gravity of mixture (air = 1.00)
  G_g = dry gas gravity
  G_0 = oil gravity
  R = surface producing gas–oil ratio in scf/stb
  M_0 = molecular weight of oil in lbm/lb-mole

The G_{mix} is then used to determine the pseudocritical properties for calculation of the compressibility factor.

If water production is quite significant, the following equation may be used [9]:

$$G_{mix} = \frac{G_g + 4584\left(\dfrac{G_0}{R} + \dfrac{1}{R_w}\right)}{1 + 132,800\left(\dfrac{G_0}{RM_0} + \dfrac{1}{18R_w}\right)} \tag{3.21}$$

where R_w = producing gas/water ratio in scf/stb

3.1.9 Tubing Performance

In a gas well, tubing performance can be defined as the behavior of the well in producing the reservoir gas through the tubing installed. At a specified surface pressure, the flowing bottomhole pressure can be calculated by using an equation for vertical flow of gas. Katz presented the equation that is simple but valid only for dry gas [10]:

$$q_g = 200,000\left(\frac{sD^5\left(P_{wf}^2 - e^sP_{wh}^2\right)}{G_g\overline{TZ}Hf(e^s - 1)}\right)^{0.5} \tag{3.22}$$

where q_g = gas flowrate in scf/d
      D = diameter of tubing in in.
      P_{wf} = bottomhole flowing pressure in psia
      P_{wh} = wellhead flowing pressure in psia
      G_g = gas gravity (air = 1.0)
      $\overline{T}$ = average temperature in °R
      $\overline{Z}$ = average gas compressibility factor
      H = vertical depth in ft.
      f = friction factor = $\{2\log[3.71/(\varepsilon/D)]\}^{-2}$
      ε = absolute pipe roughness, $\cong 0.0006$ in.

$$s = 0.0375 G_g H / \overline{TZ} \tag{3.23}$$

The average temperature used in the Equation 3.22 is simply the arithmetic average between wellhead temperature and bottomhole temperature. The gas compressibility factor $\overline{Z}$ is evaluated at the average temperature and the arithmetic average between the flowing wellhead and bottomhole pressures. This method is a trial-and-error technique, but one or two iterations is usually sufficiently accurate.

By knowing all parameters in Equation 3.22 but q_g and P_{wf}, the TPR can then be constructed. The use of TPR here is the same as discussed previously for oil wells. Figure 3.12 shows an idea of the effect of wellhead pressure on a well deliverability. A decrease in P_{wh}, thus an increase in flowrate, can be done by changing a choke/bean diameter to a bigger one.

For a specified wellhead pressure, the flowing bottomhole pressure can be estimated as a function of flowrate and tubing diameter. Equation 3.22 can be used to do this.

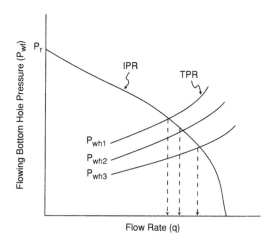

FIGURE 3.12   Effect of well head pressure on gas well deliverability.

In many cases, gas wells produce some liquids along with the gas itself. The equation for dry gas should be modified to account for the liquid content. One of the modifications presented in the literature was made by Peffer, Miller, and Hill [9]. For steady-state flow and assuming that the effects of kinetic energy are negligible, the energy balance can be arranged and written as:

$$\frac{G_g H}{53.34} = \int_{P_{wh}}^{P_{wf}} \frac{\left(\frac{P}{TZ}\right)}{F^2 + \left(\frac{P}{TZ}\right)^2} dP \tag{3.24}$$

where P_{wf} = flowing bottomhole pressure in psia
$\quad\;\; P_{wh}$ = flowing wellhead pressure in psia

$$F^2 = \frac{2666.5 f Q^2}{d^5} \tag{3.25}$$

$\quad$ d = inside diameter of tubing in in.
$\quad$ f = friction factor, dimensionless
$\quad$ Q = flowrate, MMscf/d

Evaluation of friction factor f depends on the stream of fluids in the well. For single-phase (dry gas) and fully developed turbulent flow with an absolute roughness of 0.0006 in., Cullender and Smith [8] suggested the use of:

$$F = \frac{0.10797Q}{d^{2.612}} \quad \text{for} \quad d < 4.277 \text{ in.} \tag{3.26}$$

and

$$F = \frac{0.10337Q}{d^{2.582}} \quad \text{for} \quad d > 4.277 \text{ in.} \tag{3.27}$$

When some liquids are present in the flowing stream Peffer et al. suggested the use of an apparent roughness of 0.0018 instead of using the absolute roughness of 0.0006. Also, adjustment in specific gravity of the fluids should be made by using Equation 3.20 or Equation 3.21. Applying these adjustments, the method of Cullender and Smith [8] may be used for a wide range of gas-condensate well condition.

However, whenever the Reynold's number for a specific can be calculated and pipe specifications are available, the friction factor can then be easily determined to be used for pressure loss calculations. Equation 3.28 and Equation 3.29 can be used to calculate the Reynold's number and friction factor, respectively.

$$R_e = \frac{20011 G_g Q}{\mu_g d} \tag{3.28}$$

$$\frac{1}{\sqrt{f}} = 2.28 - 4\log\left(\frac{\varepsilon}{d} + \frac{21.28}{R_e^{0.9}}\right) \tag{3.29}$$

where R_e = Reynolds number
$\quad G_g$ = gas gravity (air = 1.00)
$\quad Q$ = gas flowrate in MMscf/d
$\quad \mu_g$ = gas viscosity in cp
$\quad d$ = inside diameter of pipe in in.
$\quad \varepsilon$ = absolute roughness in in.
$\quad f$ = friction factor, dimensionless

For a well divided in equal lengths the upper half of the well has a relation:

$$37.5 G_g \frac{H}{2} = (P_{MF} - P_{WH})(I_{MF} + I_{WH}) \tag{3.30}$$

where P_{MF} = flowing pressure at midpoint in psia
$\quad P_{WH}$ = flowing wellhead pressure in psia
$\quad I$ = $(P/TZ)/[F^2 + (P/TZ)^2]$
$\quad F$ = shown in Equation 3.34
$\quad H$ = well depth in ft.

and the lower half of the well has the relation:

$$37.5 G_g H = (P_{WF} - P_{MF})(I_{WF} + I_{MF}) \tag{3.31}$$

where P_{WF} = flowing bottomhole pressure in psia.

After trial and error as previously discussed, Simpson's rule applies:

$$37.5 G_g H = \frac{P_{WF} - P_{WH}}{3}(I_{WH} + 4I_{MF} + I_{WF}) \tag{3.32}$$

3.1.10 Choke Performance

Chokes or beans are frequently installed in gas wells. These restrictions can be at the surface or at the subsurface. A surface choke is usually installed for:

1. regulating production rate
2. maintaining sufficient back pressure to avoid sand production

3. protecting surface equipment from pressure surge
4. preventing water coning
5. obeying regulatory bodies

Subsurface restrictions can be a tubing safety valve, a bottomhole choke, or a check valve. A tubing safety valve functions to stop flowstream whenever the surface control equipment is damaged or completely removed. A bottomhole choke is installed if low wellhead pressure is required or freezing of surface control equipment and lines is expected. A check valve is installed to prevent backflow of an injection well. Basically, there are two types of flow conditions: subsonic or sub-critical flow and sonic flow. The criteria to distinguish subsonic from sonic flow has been discussed previously in the section titled "Choke Performance."

For subsonic flow, the following equation given by Nind can be used to calculate the gas flowrate [11].

$$Q = 1248 CAP_u \left\{ \frac{k}{(k-1)G_g T_u} \left[\left(\frac{P_d}{P_u} \right)^{2/k} - \left(\frac{P_d}{P_u} \right)^{\frac{k+1}{k}} \right] \right\}^{0.5} \qquad (3.33)$$

where Q = gas flow rate in Mcf/d
C = discharge coefficient, ≈ 0.86
A = cross-sectional area of choke or restriction in in.^{2}
P_u = upstream pressure in psia
P_d = downstream pressure in psia
G_g = gas gravity (air = 1.00)
T_u = upstream temperature in °R
k = specific heat ratio, C_p/C_f

Equation 3.33 shows that subsonic flow is affected by upstream and downstream pressure.

In critical or sonic flow, gas flowrate depends only on upstream pressure as shown as:

$$Q = 879 CAP_u \left[\frac{k}{G_g T_u} \left(\frac{2}{k+1} \right)^{\frac{k+1}{k-1}} \right]^{0.5} \qquad (3.34)$$

where Q = flowrate in Mcf/d
C = discharge coefficient
A = choke area in in.^{2}
P_u = upstream pressure in psia
G_g = gas gravity (air = 1.00)
T_u = upstream temperature in °R
k = specific heat ratio

The discharge coefficient C can be determined using Figure 3.4 by having a knowledge of diameter ratio β and Reynold's number. Reynold's number may be calculated using the following equation:

$$R_e = \frac{20.011 Q G_g}{\mu_g d} \tag{3.35}$$

where Q = gas flowrate in Mcf/d
G_g = gas gravity (air = 1.00)
μ_g = gas viscosity, evaluated at upstream pressure and temperature in cp
d = internal diameter of pipe (not choke) in in.

Gas flow through restriction (orifice) may also be estimated using Figure 3.13. For conditions that differ from chart basis, correction factors are required. A gas throughput read from the chart must be multiplied by the proper correction factor,

$$Q = \text{Gas throughput} \times 0.0544 \sqrt{G_g T}, \text{ Mcf/d} \tag{3.36}$$

where T is the absolute operating temperature in °R.

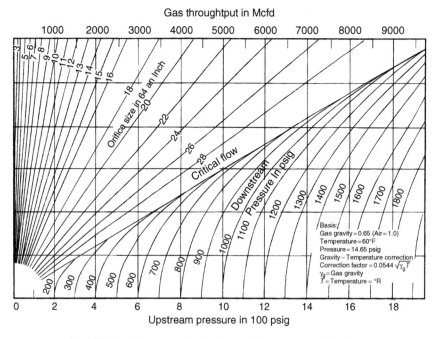

FIGURE 3.13 A correlation for gas flow through orifice [12].

3.1.11 Flowline Performance

For a single-phase gas flow, pressure-rate relation may be obtained from a known Weymouth equation:

$$Q = 433.49 \frac{T_b}{P_b} \left[\frac{(P_u^2 - P_d^2) d^{16/3}}{G_g \overline{T} \overline{Z} L} \right]^{0.5} E \tag{3.37}$$

where Q = gas flowrate in scf/d
  T_b = base temperature in °R
  P_b = base pressure in psia
  P_u = upstream pressure in psia
  P_d = downstream pressure in psia
  d = inside diameter of pipe in in.
  G_g = gas gravity (air = 1.00)
  $\overline{T}$ = average flow line temperature in °R
  $\overline{Z}$ = average gas compressibility factor
  L = pipe length in mi
  E = pipe line efficiency, fraction

or the modified Panhandle (Panhandle B) equation (for long lines):

$$Q = 737 \left(\frac{T_b}{P_b} \right)^{1.02} \left(\frac{P_u^2 - P_d^2}{\overline{T} \overline{Z} L G_g^{0.961}} \right)^{0.510} d^{2.530} E \tag{3.38}$$

with the terms and units the same as the Weymouth Equation 3.37.

The pipeline efficiency E depends on flow stream and pipeline conditions. A gas stream may contain some liquids; the higher the liquid content, the lower the line efficiency. The pipeline may be scaled, or condensate-water may accumulate in low spots in the line. Ikoku [12] presents the information about line efficiency as shown below.

Type of line	Liquid Content	E (gal/MMscf)
Dry-gas field	0.1	0.92
Casing-head gas	7.2	0.77
Gas and condensate	800	0.60

This gives an idea in estimating E for a particular condition.

Example 3.4

A gas well was flowed at a rate of 7.20 MMscf/d. The stabilized sand-face pressure at the end of the test was 1155 psia, and the current average reservoir pressure was estimated to be 1930 psia. Determine the following parameters using the modified Mishra-Caudle [4] method:

1. AOFP at current conditions ($P_r = 1{,}930$ psia)
2. deliverability at $P_{wf} = 1{,}000$ psia
3. AOFP at a future average pressure $P_{rf} = 1{,}600$ psia
4. deliverability at a future $P_{wf_f} = 1{,}155$ psia

Solution

The readers may refer to the original paper [4]. Here the modified one is given

a. For a pressure less than 2100 psia, Equation 3.12 is used:

$$\frac{q}{q_{max}} = \frac{5}{4}\left[1 - 5^{\left(\frac{P_{wf}^2}{P_r^2} - 1\right)}\right]$$

$$AOFP = q_{max} = \frac{7.2\ \text{MMscf/d}}{\frac{5}{4}\left[1 - 5^{\left(\frac{1155^2}{1930^2} - 1\right)}\right]} = 8.94\ \text{MMscf/d}$$

b. $q = 8.94 \times 10^6 \times \dfrac{5}{4}\left[1 - 5^{\left(\frac{1000^2}{1930^2} - 1\right)}\right]$

$= 7.73\ \text{MMscf/d}$

c. $\dfrac{(AOFP)_f}{(AOFP)_p} = \dfrac{5}{3}\left[1 - 0.4^{\frac{P_{rf}^2}{P_{rp}^2}}\right]$

$(AOFP)_f = 8.94 \times 10^6 \times \frac{5}{3}\left(1 - 0.4^{\frac{1600^2}{1930^2}}\right)$

$= 6.96\ \text{MMscf/d}$

d. Again use Equation 3.12:

$$q_f = (AOFP) \times \frac{5}{4}\left[1 - 5^{\left(\frac{P_{wf}^2}{P_r^2}\right)_f - 1}\right]$$

$$= 6.96 \times 10^6 \times \frac{5}{4}\left[1 - 5^{\left(\frac{1155}{1600}\right)^2 - 1}\right]$$

$$= 4.675\ \text{MMscf/d}$$

Example 3.5

The following data are obtained from a producing gas well:

Gas gravity $= 0.65$
Well depth (vertical) $= 6000$ ft.
Wellhead temperature $= 570°R$
Formation temperature $= 630°R$
Flowing wellhead pressure $= 1165$ psia
Flowrate $= 10.0$ MMscf/d
Tubing ID $= 2.441$ in.
Pseudo-critical temperature $= 374°R$
Pseudo-critical pressure $= 669$ psia
Absolute roughness $= 0.00065$

Calculate the following bottomhole pressure using Equation 3.22.

Solution

Rearranging Equation 3.22, we obtain:

$$P_{wf} = \left(e^s P_{wh}^2 + \frac{q_g^2}{4 \times 10^{10}} \times \frac{G_g \overline{TZ} H f (e^s - 1)}{S D^5} \right)^{0.5}$$

Assume $P_{wf} = 2000$ psia:

$$\overline{P} = \frac{2000 + 1165}{2} = 1582.5 \text{ psia}$$

$$T = \frac{570 + 630}{2} = 600°R$$

$$\left. \begin{array}{l} pP_r = \dfrac{1582.5}{669} = 2.365 \\[2mm] PT_r = \dfrac{600}{374} = 1.064 \end{array} \right\} \rightarrow \overline{Z} = 0.848$$

$$S = 0.0375(0.65)(6,000)/(600)(0.848) = 0.287$$

$$f = \left\{ 2\log \left[371 \Big/ \left(\frac{0.00065}{2.441} \right) \right] \right\}^{-2}$$

$$P_{wf} = \left\{ e^{0.287}(1,165)^2 + \frac{(10 \times 10^6)^2}{4 \times 10^{10}} \times \frac{(0.65)(600)(0.848)(6,000)}{(0.287)(2.441)^5} \right.$$

$$\left. \times \left[0.0146(e^{0.287} - 1) \right] \right\}^{0.5}$$

$$= 1,666 \text{ psia} \ (\ll 2,000 \text{ psia assumed})$$

Assume 2nd $P_{wf} = 1{,}666$ psia:

$$\overline{P} = \frac{1{,}666 + 1{,}165}{2} = 141.5 \text{ psia}$$

$$\left.\begin{array}{l} pP_r = \dfrac{141.5}{669} = 2.116 \\[2mm] pT_r = 1.604 \end{array}\right\} \rightarrow \overline{Z} = 0.857$$

$$S = 0.0375(0.65)(6{,}000)/(600)(0.857) = 0.284$$

$$P_{wf} = 1666.7 \text{ psia } (\cong 1{,}666 \text{ psia assumed})$$

Example 3.6

Suppose the gas well in Example 3.5 is produced through a flowline of 2.5 in. in diameter and 1250 ft long. The average operating temperature is 100°F. Additional data given are specific heat ratio, $k = 1.3$, and the estimated gas viscosity at the operating condition, $\mu_g = 0.0131$ cp.

Find the positive choke size required for critical condition, and the pressure at downstream of the flowline.

Solution

1. For a critical flow, Equation 3.34 can be used.

$$Q = 879 \times C \times A \times P_u \left[\frac{k}{G_g T_u} \left(\frac{2}{k+1} \right)^{\frac{k+1}{k-1}} \right]^{0.5}$$

$$= 879 \times C \times \frac{\pi}{4} d_{ch}^2 \times 1165 \times \left[\frac{1.3}{0.65 \times 570} \left(\frac{2}{1.3+1} \right)^{\frac{1.3+1}{1.3-1}} \right]^{0.5}$$

$$= 27{,}881 C d_{ch}^2 \ (\text{Mscf/d})$$

$$R_e = \frac{20 \times 10{,}100 \times 0.65}{0.013 \times 2.5} = 4.01 \times 10^6$$

Assume a diameter choke, such that $d_{ch} = 0.7$ in. Use Figure 3.4 to find coefficient C:

$$\beta = \frac{d_{ch}}{d_{pipe}} = \frac{0.7}{2.5} = 0.28$$

For $R_e = 4.01 \times 10^6$ and $\beta = 0.28$, $C = 0.998$, so

$$Q = 27{,}881 \times 0.998 \times (0.7)^2 = 13.63 \text{ MMscf/d}$$

If we assume another choke size, let $d_{ch} = 0.6$ in.:

$$\beta = \frac{0.6}{2.5} = 0.24 \rightarrow C = 0.997$$

$$Q = 27{,}881 \times 0.997 \times (0.6)^2 = 10.0 \text{ MMscf/d}$$

2. For a critical flow, we may assume $(P_d)_{choke} = 0.5(P_u)_{choke}$. Let's take $T_b = 530°R$ and $P_b = 15$ psia. Then using Equation 3.37 with $E = 0.92$,

$$P_d^2 = P_u^2 - \left(\frac{Q/E}{433.4\,T_b/P_b}\right)^2 \times \frac{G_g \overline{T} Z L}{d^{16/3}}$$

$$= \left(\frac{1165}{2}\right)^2 - \left(\frac{10^7/0.92}{15{,}317}\right)^2 \times \frac{(0.65)(560)(\overline{Z})\left(\dfrac{1250}{5280}\right)}{(2.5)^{16/3}}$$

$$= 339{,}306 - 327{,}422\overline{Z}$$

Assume that $P_d = 200$ psia to find $\overline{Z}$ such that:

$$\overline{P} = \frac{200 + (0.5 \times 1165)}{2} = 391.2 \text{ psia}$$

$$\left.\begin{array}{l} pP_r = \dfrac{391.25}{669} = 0.585 \\[2mm] pT_r = \dfrac{560}{374} = 1.497 \end{array}\right\} \rightarrow \overline{Z} = 0.939$$

$$P_d^2 = 339{,}306 - 327{,}422 \times 0.939$$
$$P_d = 178.5 \text{ psia}(\neq P_{d_{assumed}} = 200 \text{ psia})$$

Assume now that $P_d = 178.5$ psia such that:

$$\overline{P} = \frac{178.5 + (0.5 \times 1165)}{2} = 380.5 \text{ psia}$$

$$\left.\begin{array}{l} pP_r = \dfrac{380.5}{669} = 0.569 \\[2mm] pT_r = 1.497 \end{array}\right\} \rightarrow \sim\overline{Z} = 0.940$$

$$P_d^2 = 339{,}306 - 327{,}422 \times 0.940$$
$$P_d = 177.5 \text{ psia}(\cong P_{d_{assumed}} = 178.5 \text{ psia})$$

3.2 TWO-PHASE FLOW PERFORMANCE

3.2.1 Two-Phase Inflow Performance

When a reservoir pressure is below the bubble point pressure, the simple equation of inflow performance (e.g., the productivity index is constant) is no longer valid, because at this condition the oil flowrate will decline much faster at increasing drawdown than would be predicted by Equation 3.1a or Equation 3.2. An illustrative comparison of the two types of IPR is shown in Figure 3.14.

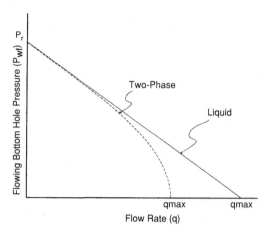

FIGURE 3.14   Illustrative comparison of liquid and two-phase IPR curves.

3.2.2 Vogel's Method

The well-known inflow performance equation for two-phase flow has been proposed by Vogel [13]. The equation is,

$$\frac{q_0}{q_{max}} = 1 - 0.2\frac{P_{wf}}{P_r} - 0.8\left(\frac{P_{wf}}{P_r}\right)^2 \qquad (3.39)$$

which fits a general dimensionless IPR shown in Figure 3.15. The reference curve and Equation 3.39 is valid for solution-gas drive reservoir with reservoir pressures below the bubble point. The formation skin effect is not taken into account. The method is originally developed with the flowing efficiency FE = 1.0. However, for a given well with any FE known, Equation 3.39 or the reference curve may be used to generate the IPR curve.

For reservoir pressures above the bubble point but with flowing pressures below the bubble point, the constant J equation and the Vogel equation can be combined to estimate the IPR curves. The equation is

$$q_0 = q_b + \left(q_{max} - q_b\right)\left[1 - 0.2\frac{P_{wf}}{P_r} - 0.8\left(\frac{P_{wf}}{P_r}\right)^2\right] \qquad (3.40)$$

The maximum flowrate q_{max} is calculated using the following equation:

where q_0 = oil flowrate in stb/d

$\quad q_{max}$ = the theoretical maximum flowrate when $P_{wf} = 0$ in stb/d

$\quad q_b$ = oil flowrate at $P_{wf} = P_b$ in stb/d

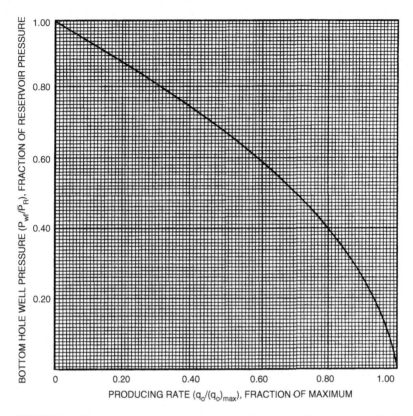

FIGURE 3.15 A general dimensionless IPR for solution gas drive reservoirs [13].

P_b = bubble point pressure in psia
P_{wf} = flowing bottomhole pressure in psia
P_r = average reservoir pressure in psia

$$q_{max} = q_b + \frac{J \times P_b}{1.8} \qquad (3.41)$$

The productivity index J is determined based on the following bottomhole pressure of the test:

1. For $(P_{wf})_{test} > P_b$, then

$$J = \frac{(q_0)_{test}}{P_r - (P_{wf})_{test}} \qquad (3.42)$$

2. For $(P_{wf})_{test} < P_b$, then

$$J = \frac{(q_0)_{test}}{P_r - P_b + (P_b \times M/1.8)} \qquad (3.43)$$

where $M = (1 - 0.2(P_{wf})/P_b) - 0.8(P_{wf}/P_r)^2$

$P_{wf} = (P_{wf})_{test}$

The q_b is calculated using Equation 3.2 with $P_{wf} = P_b$

3.2.3 Fetkovich Method

Analyzing isochronal and flow–afterflow multipoint back-pressure tests conducted on oil wells, Fetkovich [14] found that back-pressure curves for oil wells followed the same form as for gas wells; that is

$$q_0 = J'_0 (P_r^2 - P_{wf}^2)^n \tag{3.44}$$

where J'_0 = back-pressure curve coefficient, stb/d/(psia)^{2n}

$\quad\quad n$ = back-pressure curve exponent or exponent
of inflow performance curve

The plot of q_0 versus $(P_r^2 - P_{wf}^2)$ on log-log paper is considered as good as was obtained from gas well back-pressure tests. Conducting a multipoint back-pressure test on a well, Equation 3.44 can be used to predict the IPR curve for the well.

Figure 3.16 shows the comparison of IPR's for liquid, gas and two-phase (gas and liquid). Fetkovich reported that Vogel's equation yields $n = 1.24$ (Figure 3.17).

For reservoir pressures above bubble point pressures, the inflow performance curves can be constructed using the following equation:

$$q_0 = J'_0 (P_b^2 - P_{wf}^2)^n + J(P_r - P_b) \tag{3.45}$$

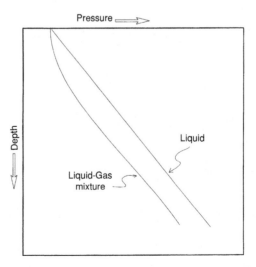

FIGURE 3.16   Pressure gradients of flowing liquid and liquid–gas mixture.

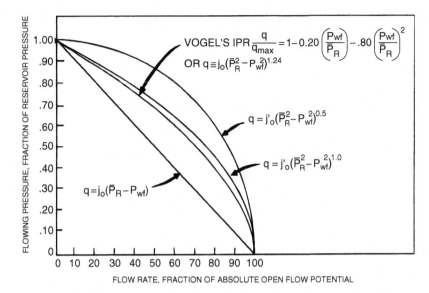

FIGURE 3.17   Inflow performance relationship for various flow equations [14].

The maximum flowrate of a well can be determined using the following equation:

$$q_{max} = q_0 \left[1 - \left(\frac{P_{wf}}{P_r} \right)^2 \right]^n \tag{3.46}$$

3.2.4 Modified Standing's Method

Vogel's reference curve is originally derived for undamaged or unformulated wells. In other words, the curve is only valid for wells with skin factor $s = 0$. Later, Standing [15] presented a set of companion curves that can be used to predict IPR curves for damaged or stimulated wells. His method is based on the definition of single-phase flow efficiency. In fact, solution-gas drive reservoirs producing oil at $P_{wf} < P_b$ and/or $P_r < P_b$ have inflow performance of two-phase flow. The IPR of this type of reservoirs have been shown to have quadratic forms as suggested by Vogel and Fetkovich.

Camacho and Raghhavan [16] found that Standing's definition of flow efficiency is incorrect to be used in two-phase flow behavior. It is suggested that the definition of flow efficiency must also reflect the quadratic form of the inflow performance equation. This is expressed by

$$FE = \frac{(1 + VP'_{wf}/P_r)(1 - P'_{wf}/P_r)}{(1 + VP_{wf}/P_r)(1 - P_{wf}/P_r)} \tag{3.47}$$

where FE = flowing efficiency
$\quad$ V = quadratic curve factor (V = 0.8 for Vogel
$\qquad$ and V = 1.0 for Fetkovich)
$\quad$ P'_{wf} = ideal flowing bottomhole pressure (e.g.,
$\qquad$ when skin factor s = 0) in psia
$\quad$ P_{wf} = actual flowing bottomhole pressure in psia
$\quad$ P_r = average reservoir pressure in psia

The flowrate when FE $\neq$ 1.0 can be calculated using

$$\frac{q_0}{q_{max}^{FE=1.0}} = FE\left[1 + V\left(\frac{P_{wf}}{P_r}\right)\right]\left[1 - \left(\frac{P_{wf}}{P_r}\right)\right] \tag{3.48}$$

where $q_{max}^{FE=1.0}$ = maximum flow rate for undamaged/unstimulated
$\qquad$ well or when FE = 1.0

3.2.5 Predicting Future IPR

Predicting future well deliverability is frequently needed in most oil fields. Some of the many reasons are:

1. to estimate when the choke should be changed or adjusted to maintain the production rate
2. to predict well capability and evaluate if the tubing has to be changed
3. to do planning for selecting future artificial lift methods
4. to do planning for reservoir pressure maintenance or secondary recovery projects

Some prediction methods available in the literature are discussed.

3.2.5.1 Standing's Method

The method has been developed based on Vogel's equation, the definition of productivity index, and the assumption that the fluid saturation is to be the same everywhere in the reservoir [17]. Three basic equations presented are:

$$J_p^* = \frac{1.8(q_{max})_p}{(P_r)_p} \tag{3.49}$$

$$J_f^* = J_p^* \left\{ \frac{[K_{r0}/(\mu_0 B_0)]_f}{[k_{r0}/(\mu_0 B_0)]_p} \right\} \tag{3.50}$$

$$(q_{max})_f = (J_f^* \cdot P_{rf})/1.8 \tag{3.51}$$

where J^* = productivity index at zero drawdown in stb/d/psi
$\quad$ k_{r0} = relative permeability to oil, fraction
$\quad$ μ_0 = oil viscosity in cp
$\quad$ B_0 = oil formation volume factor in bbl/stb

and subscript p and f refer to present and future conditions, respectively. The relative permeability to oil is at corresponding oil saturation in the reservoir. A method for determining oil saturation and k_{r0} may be found in the book *Reservoir Engineering*. In the example presented by Standing, Tarner's method was used and k_{r0} was evaluated using Corey-type relationship:

$$K_{r0} = \left(\frac{S_0 - S_{or}}{1 - S_{wc} - S_{or}} \right)^4 \qquad (3.52)$$

where S_{or} = residual oil saturation, fraction
S_{wc} = connate water saturation, fraction

3.2.5.2 Combined Fetkovich-Vogel Method

Fetkovich suggested that a future well deliverability may be estimated by the relation $J'_f / J'_p = (P_r)_f / (P_r)_p$. Recalling Equation 3.44 and taking $n = 1$, we can write

$$\frac{(q_{max})_f}{(q_{max})_p} = \left[\frac{(P_r)_f}{(P_r)_p} \right]^3 \qquad (3.53a)$$

or

$$(q_{max})_f = (q_{max})_p \times \frac{(P_r)_f^3}{(P_r)_p^3} \qquad (3.53b)$$

After calculating the maximum flowrate using Equation 3.53b, the inflow performance curve into the future can be constructed using Vogel's equation. This method is valid only for undamaged wells.

3.2.5.3 Unified Method

Kelkar and Cox proposed a new method for predicting future IPR [18]. This method is a result of unification of some methods discussed previously. The relationship suggested can be applied to any of the reference methods. Two sets of data points (each at different average reservoir pressure) are required to predict the future inflow performance curve. The procedure is as follows:

1. Calculate the maximum flowrate (q_{max} or Q_{max}) for both tests conducted using the reference method (Vogel, Fetkovich or Modified Standing).
2. Calculate J*:

$$J_1^* = \frac{(q_{max})_1}{(P_r)_1} \quad \text{and} \quad J_2^* = \frac{(q_{max})_2}{(P_r)_2} \qquad (3.54)$$

3. Determine constants A' and B' as

$$A' = \frac{J_1^* - J_2^*}{(P_r^2)_1 - (P_r^2)_2} \tag{3.55}$$

$$B' = \frac{\dfrac{J_1^*}{(P_r^2)_1} - \dfrac{J_2^*}{(P_r^2)_2}}{\dfrac{1}{(P_r^2)_1} - \dfrac{1}{(P_r^2)_2}} \tag{3.56}$$

4. Calculate the maximum flowrate of the corresponding future pressure $(P_r)_f$

$$(q_{max})_f = A'(P_r^3)_f + B'(P_r)_f \tag{3.57}$$

5. Construct the future inflow performance curve using the reference inflow performance equation used in Step 1 above for reservoir pressure $(P_r)_f$ and the maximum flowrate calculated in Step 4.

3.2.6 Tubing Performance

The problem of simultaneous flow of oil, gas, and water through the vertical tubing of an oil well is complex. The fluid is a compressible mixture, its density increasing with depth. The gradient line has a distinct curvature (see Figure 3.16). Along the gradient line of a given well, different flow regimes occur, which may range from a mist flow in the region of low pressures to a single-phase flow at the pressures when all gas is in the solution.

The knowledge of tubing performance of flowing wells is important for efficient operations. Present and future performance of the wells may be evaluated. This may suggest changes in operating practices and equipment to prolong the flowing life of a well. Figures 3.18 and 3.19 show the idea of the effects of tubing size and a change in IPR on a well performance, respectively.

For a given wellhead pressure, flowing bottomhole pressure varies with production rate. Plotting these two flowing parameters on a Cartesian coordinate will give a curve called *tubing performance relationship* (TPR). By plotting the TPR and IPR of an oil well on a graph paper, the stabilized production capacity of the well is represented by the intersection of the two curves (see Figure 3.18).

To construct a TPR curve for a given well, the fluids and well geometry data should be available. Chapter 2 "Flow of Fluids" provides some good multiphase flow correlations that can be used. In cases where these data and accessibility to computer are limited, a graphical flowing gradients correlation is needed. In fact, many improved graphical correlations covering a broad range of field conditions are available in the literature. The readers may refer to Brown [19] to get a complete set of flowing gradient curves.

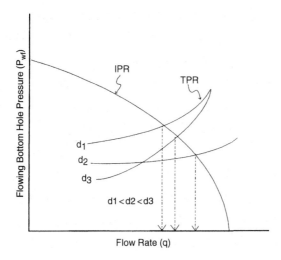

FIGURE 3.18   Effect of tubing size on oil well deliverability.

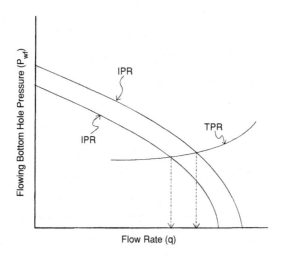

FIGURE 3.19   Effect of changes in inflow performance on oil well productivity.

Sometimes a different company has a different set of curves. Although the best correlation is available, a particular field condition might have specific well characteristics such as salt or asphaltene deposition in the tubing or severe emulsification that may bring about a higher pressure drop than would be estimated from existing graphical correlations. The discrepancy might be used for analyzing the wells. So, it is recommended that a good multiphase flow equation be used instead of a graphical correlation.

For convenience, however, a set of working curves developed by Gilbert [20] is presented for illustrating their use in solving well performance problems. The curves are shown in Figures 3.20 through Figure 3.29, and available only for small flowrate (50 to 600 stb/d) and tubing sizes of 1.66, 1.90, 2.375, 2.875, and 3.50 in.

The procedure how to use these flowing gradient curves in production engineering problems is given in the example later.

3.2.7 Use of Vertical Pressure Gradients

In the preceding section of single-phase flow performances, the functions of vertical flow performance curves have been discussed. The following is a more detailed discussion on the applications of pressure gradients in analyses of flowing well performance. Accurate well test data, obtained under stabilized flow conditions, are needed for such analyses.

3.2.7.1 Subsurface Data

With flowing tubing known and well test data available, the flowing bottomhole pressure for a given rate of production can be determined by calculating the flowing pressure gradient to the bottom of the well.

If the static bottomhole pressure is known, the productivity index of the well can be determined from one production rate by determining the flowing bottomhole pressure for the rate of production.

If only surface data are available, the productivity index of the well may be estimated by determining the flowing bottomhole pressures for two or more rates of production.

3.2.7.2 Tubing Size

As stated, the size of the tubing is one of the important parameters affecting the pressure gradients. For low velocity, the slippage of gas by the liquid contributes to the pressure losses. For high velocities, friction becomes the controlling factor. Between these two extremes, there is a range of velocities giving the optimum gradient at the inlet of tubing in the bottom of the well.

If the future range of expected rates and gas/oil ratios can be estimated, selection of the tubing size can be made, which would assure operation within the efficient range of the gradients, with the resulting increase in the flowing life of the well. Such selection can be made by calculating gradients for different tubing sizes for a given set of conditions.

3.2.8 Water Content

As the gas/oil ratio decreases, the flowing gradients, of course, increase, other conditions being equal. This is clearly illustrated in Figure 3.20 for example. As the water content of the produced fluid increases, the overall gas/liquid ratio decreases. If the future behavior of the water content increase can be estimated, future behavior of the well can be evaluated.

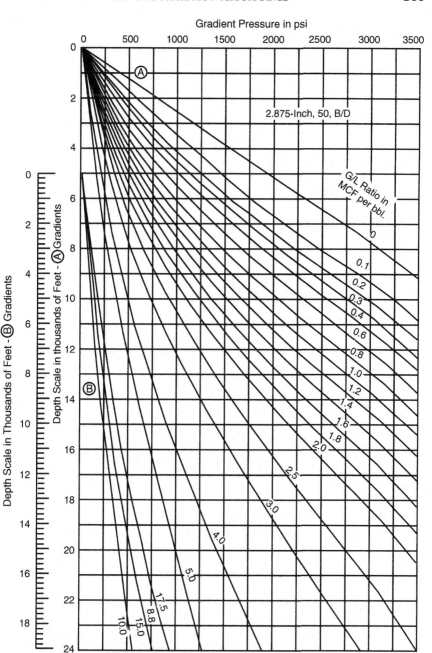

FIGURE 3.20 Flowing pressure gradients for 2.875-in. tubing with rate of 50 bpd.

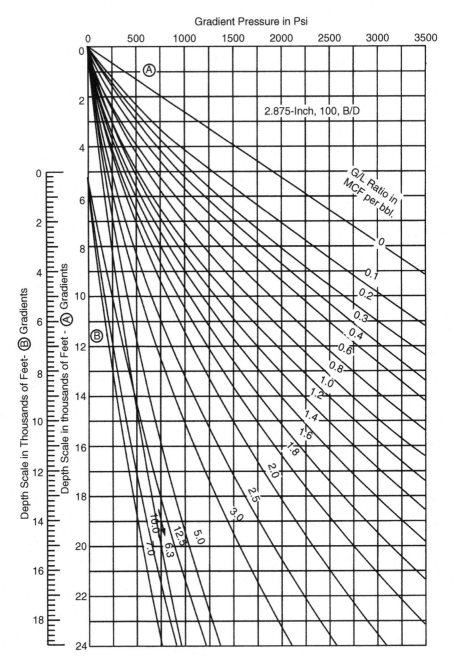

FIGURE 3.21   Flowing pressure gradients for 2.875-in. tubing with rate of 100 bpd.

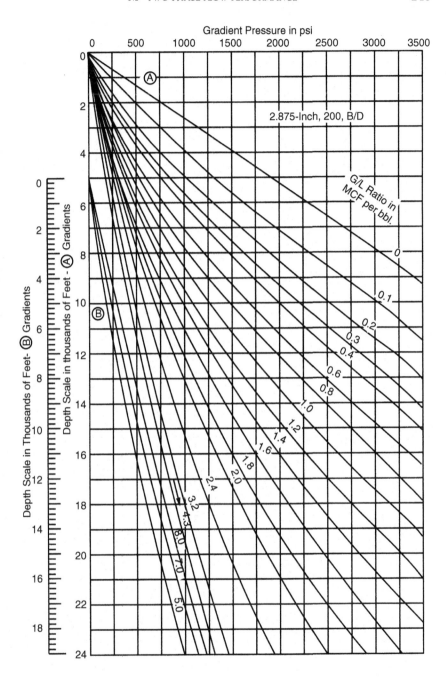

FIGURE 3.22 Flowing pressure gradients for 2.875-in. tubing with rate of 200 bpd.

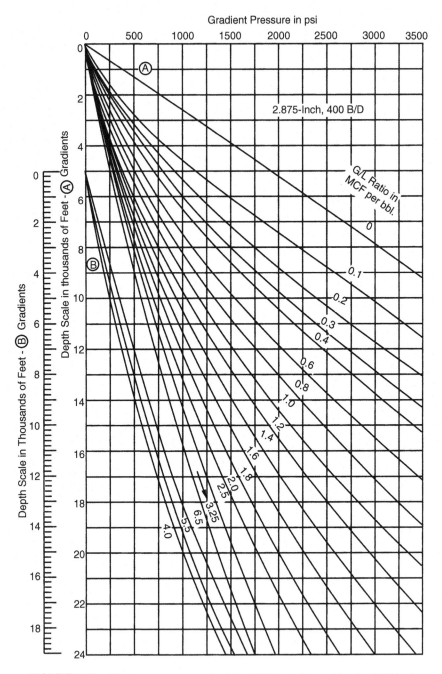

FIGURE 3.23   Flowing pressure gradients for 2.875-in. tubing with rate of 400 bpd.

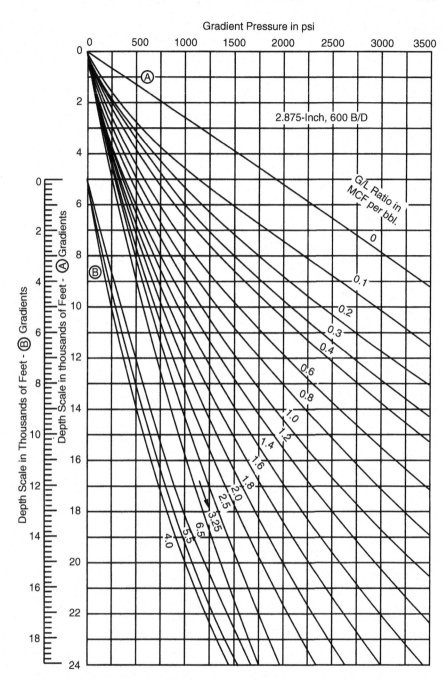

FIGURE 3.24 Flowing pressure gradients for 2.875-in. tubing with rate of 600 bpd.

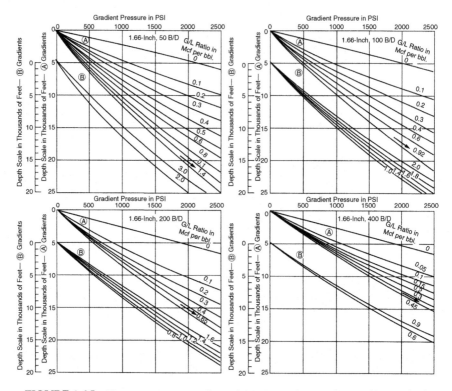

FIGURE 3.25 Flowing pressure gradients for 1.66-in. tubing with rate of 50–400 bpd.

3.2.9 Wellhead Pressure

To a degree, the wellhead pressure is controllable because it depends, among other things, on the size, length, and geometry of flow lines and on the separator pressure. At the same time, the wellhead pressure has a marked effect on the slope characteristics of the gradient because of the question of density which is involved. For a given reduction of the wellhead pressure, the reduction of the bottomhole flowing pressure may be substantially higher. This is particularly true in case of high-density flow.

The reduction of the bottomhole flowing pressure should result in increase of production. It can be seen from the above that the degree of this increase cannot be estimated from the surface data alone. It can be estimated if information is available also on the flowing gradient and the productivity index of the well.

3.2.10 Predicting the Flowing Life

The natural flow of an oil well continues as long as there exists a proper balance between two conflicting pressure requirements at the bottom of

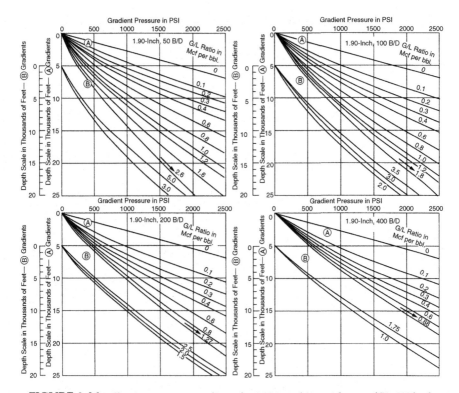

FIGURE 3.26 Flowing pressure gradients for 1.90-in. tubing with rate of 50–400 bpd.

the well. First, this pressure must be sufficiently high to sustain the vertical lift. Second, it must be sufficiently low to create a pressure differential that permits reservoir fluids to enter the well.

This balance may be destroyed by either one or a combination of two sets of conditions:

1. increase in flowing pressure gradients for any of the reasons mentioned above, increases in the lift pressure requirements above the point needed for maintaining the pressure differential with the available reservoir energy.
2. the declining reservoir energy is not able to maintain this differential for the required vertical lift pressure.

In either case, the natural flow of the well either declines to an uneconomical rate or ceases completely.

The uses of the flowing pressure gradients discussed above may be applied to estimating the length of the flowing life of a well. Additional information needed are the estimate of the static bottomhole pressure of water encroachment, and of gas/oil ratio behavior at different future stages

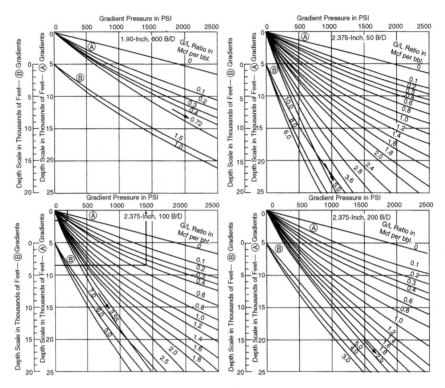

FIGURE 3.27   Flowing pressure gradients for 1.90-in. tubing with rate of 600 bpd and 2.375-in. tubing with rate of 50–200 bpd.

of the cumulative production of the well. Graphic methods have been developed for making such estimates.

3.2.11 Choke Performance

The reasons involved in installing chokes in oil and gas fields have been mentioned before. The graphical correlations and empirical or semiempirical equations used for single-phase oil or gas are not valid for two-phase conditions.

The correlations for multiphase flow through chokes have been published, but not one of them gives satisfactory results for all ranges of operating conditions (flow parameters). Theoretically, the correlations are developed with the assumption that the simultaneous flow of liquid and gas is under critical flow conditions so that when an oil well choke is installed fluctuations in line pressure is gradually increased, there will be no change in either the flowrate or the upstream pressure until the critical–subcritical flow boundary ($P_{downstream} \cong 0.5 - 0.55\, P_{upstream}$) is reached.

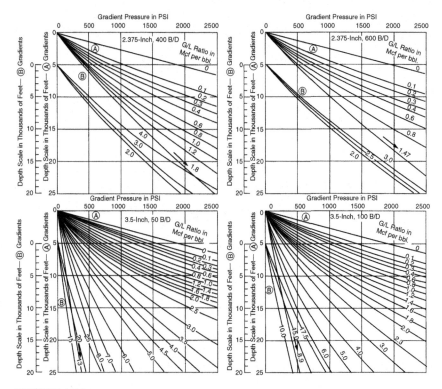

FIGURE 3.28 Flowing pressure gradients for 2.375-in. tubing with rate of 400–600 bpd and 3.5-in. tubing with rate of 50–100 bpd.

In more oil fields, the most popular correlations are of Gilbert [20] and Poetmann and Beck [21]. Ashford [22] also developed a correlation for multiphase flow through chokes.

3.2.12 Gilbert's Correlation

The equation developed to estimate a parameter of fluid flow through the orifice is:

$$P_{wh} = \frac{435 R^{0.546} q}{d^{1.89}} \tag{3.58}$$

where P_{wh} = wellhead pressure in psig
R = gas/liquid ratio in Mscf/stb
q = gross liquid rate in stb/d
d = choke (bean) size in 1/64 in.

This equation is derived using regularly reported daily individual well production data from Ten Section Field in California. Gilbert noted that an error of $\frac{1}{128}$ in. in bean size can give an error of 5 to 20% in pressure

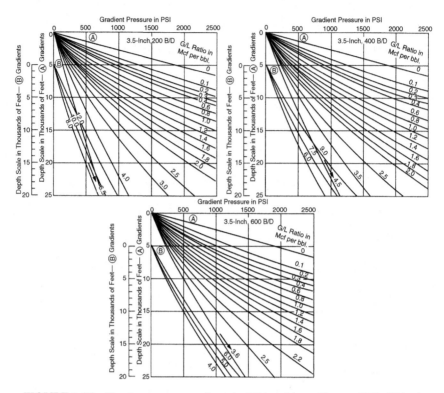

FIGURE 3.29   Flowing pressure gradients for 3.5-in. tubing with rate of 200–600 bpd.

estimates. In the type of formula used, it is assumed that actual mixture velocities through the bean exceed the speed of sound, for which condition the downstream, or flow line, pressure has no effect upon the tubing pressure. Thus, the equation applies for tubing head pressure of at least 70% greater than the flow line pressure.

3.2.13   Poetmann–Beck's Correlation

F. H. Poetmann and R. L. Beck [21] developed charts for estimating flow of oil and gas through chokes. The charts shown in Figures 3.30 through 3.32, relate the variables of gas/oil ratio, oil production rate, tubing pressure, and choke size. With three of the variables known, the fourth can be determined. The charts are valid only under the following conditions:

1. The flow is a simultaneous, two-phase flow of oil and gas. The charts are not valid if water is present.
2. The flow through the choke is at the critical flow conditions; that is, at acoustic velocity. This occurs when the downstream (flow line) pressure

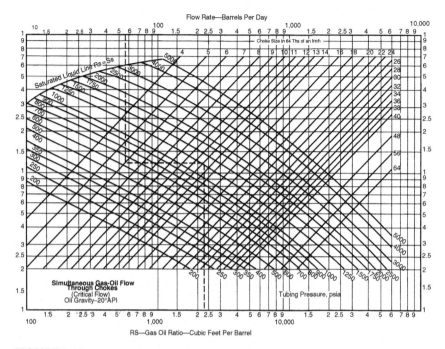

FIGURE 3.30 A correlation for two-phase flow through chokes with oil gravity of 20° API.

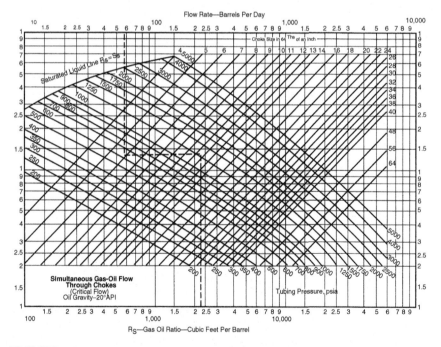

FIGURE 3.31 A correlation for two-phase flow through chokes with oil gravity of 30° API.

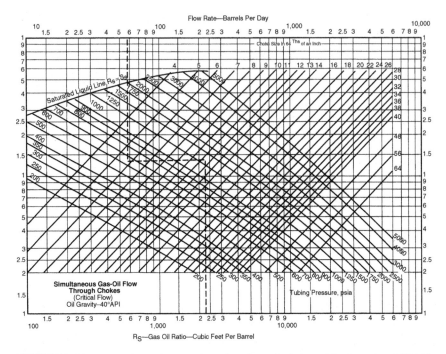

FIGURE 3.32   A correlation for two-phase flow through chokes with oil gravity of 40°API.

is 0.55 or less of the upstream (tubing) pressure. Under such conditions the rate of flow is not affected by downstream pressure.

Actually, this last limitation is of small practical significance since chokes are usually selected to operate at critical flow conditions so that the well's rate of flow is not affected by changes in flowline pressure.

The manner of use of these charts is as follows:

1. The 20°, 30°, and 40° API gravity charts can be used for gravity ranges of 15° to 24°, 25° to 34°, and 35° and up, respectively.
2. When the starting point is the bottom scale, a vertical line is drawn to intersection with tubing pressure curve, then a horizontal line to intersection with the choke size, and then vertical line to the upper scale. Reverse procedure is used when the upper scale is the point of beginning.
3. Performance of a given size choke for a given gas/oil ratio can be plotted. Such a plot would show relationship between different rates and corresponding tubing pressures for a given choke.
4. Free gas in tubing can be estimated by use of charts. For instance, in case of Figure 3.32 for 2250 ft.[3]/bbl gas/oil ratio, 1265 psi tubing pressure and 6/64 in. choke, the rate was found to be 60 bbl/d. For 1265 psi

tubing pressure the solution gas is 310 ft.^{3}/bbl. Therefore, the free gas is $2250 - 310 = 1940$ ft.^{3}/bbl.

The results obtained by use of these charts compared very favorably with observed data obtained on 108 wells covering a wide range of conditions.

3.2.14 Ashford's Correlations

Ashford [22] developed a model for multiphase flow through a choke by applying the polytropic expansion theory. The final form of his equations is:

$$q_0 = 1.53 \frac{Cd^2 P_u}{(B_0 + WOR)^{1/2}} \times \frac{[(T_u Z_u (R - R_s) + 151 P_u)]^{1/2}}{T_u Z_u (R - R_s) + 111 P_1}$$

$$\times \frac{[G_0 + 0.000217 G_g R_s + (WOR) G_w]^{1/2}}{G_0 + 0.000217 G_g R + (WOR) G_w} \qquad (3.59)$$

Where q_0 = oil flowrate in stb/d

    C = orifice discharge coefficient

    d = choke diameter in 1/64 in.

    P_u = upstream choke pressure in psia

    B_0 = oil formation volume factor in bbl/stb

  WOR = water/oil ratio

    R = producing gas-oil ratio at standard condition in scf/stb

    R_s = solution gas-oil ratio at choke conditions in scf/stb

    T_u = upstream choke temperature in °R

    Z_u = gas compressibility factor evaluated at upstream conditions

    G_g = gas gravity (air = 1.00)

    G_0 = oil gravity (water = 1.00)

    G_w = formation water gravity (water = 1.00)

Ashford stated that once C (discharge coefficient) has been fairly well defined for a given production province or operation the Equation 3.59 may be used in a conventional manner to evaluate:

1. flowrates arising from changes in choke sizes.
2. wellhead pressures arising from changes in choke sizes.
3. choke sizes necessary to achieve a given wellhead pressure for a known liquid rate.

If C is unknown, a value of 1.0 may be used to obtain a reasonable estimate of choke performance. Later, based on an extensive study, Sachdeva et al., [23] recommended that $C = 0.75$ be used for a choke configuration involving an elbow upstream from the choke and $C = 0.85$ be used for a choke free of upstream disturbances.

3.2.15 Flowline Performance

Understanding the behavior of multiphase fluids flow in horizontal or inclined pipe is important because the efficiency of a producing system is accomplished by analyzing all components through which the fluids flow. In the analysis a flowline may be considered as a restriction because higher pressure loss resulted. For instance, for a given set of fluids data 2.5-in. line causes higher pressure loss when compared with 3 and 4-in. line, so one tends to take the larger size to produce more oil. The diameter, however, should not be oversized because additional slugging and heading may occur. Some operators just add a parallel line instead of replacing the current line with a larger size. It should be remembered that production capacity, pipes availability, separator pressure, and other constraints may be involved in judging a final design of producing system.

The knowledge of pressure flowrate relationship is very useful in designing an efficient flowing system. The procedure used to generate a flowline performance for a given set of fluids data and a given diameter of the pipe is similar with the one for single-phase. No pressure loss correlations for horizontal or inclined pipes are given. The reader can, however, find some good correlations in Chapter 2 titled "Flow of Fluids."

Example 3.7

Consider an oil reservoir producing at an average reservoir pressure below $P_r = 2400$ psia, which is the bubble point pressure. A single-point flow test conducted on a well at stabilized condition resulted in $q = 500$ stb/d and $P_{wf} = 250$ psi. Measured GOR is 400 scf/stb. The well of total depth of 7000 ft. is produced through $2\frac{3}{8}$-in. tubing. Find (1) the maximum rate possible assuming $P_{wh} = 200$ psi, and (2) the productivity index for the condition corresponding to the maximum possible flowrate.

Solution (assume FE = 1.0)

1.

1. Using Vogel's equation, calculate q_{max}

$$\frac{500}{q_{max}} = 1 - 0.2\left(\frac{250}{2,400}\right) - 0.8\left(\frac{250}{2400}\right)^2$$

$q_{max} = 515$ stb/d($=$ oil AOFP)

2. Choose some values of P_{wf} and determine corresponding q's using Vogel's equation

$$q = 515\left[1 - 0.2\left(\frac{P_{wf}}{2400}\right) - 0.8\left(\frac{P_{wf}}{2400}\right)^2\right]$$

P_{wf}, psi	q, stb/d
500	476
800	435
1200	361
1500	290
2000	143
2400	0

Plot P_{wf} versus q to get the IPR curve (see Figure 3.33).

3. Let's select Gilbert's working curves shown in Figures 3.26 and 3.27 for $2\frac{3}{8}$-in. tubing and choose the curves with rates of 100, 200 and 400 stb/d.

4. Determine the equivalent length depth of $P_{wf} = 200$ psi. This is done by tracing down a vertical line through pressure point of 200 psi at zero depth until the line of GLR = 400 scf/stb is found and read the depth. This is the equivalent depth. Add this equivalent depth to the actual well depth. Then find the

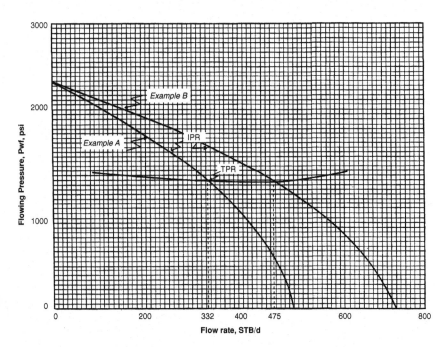

FIGURE 3.33 Pressure rate relationship for Example 3.7.

pressure at this total equivalent depth for rate and GLR given. For all three rates chosen, we can obtain

q, stb/d	P_{wf}, psi
100	11410
200	11370
400	11310
(600)	(1400)

5. Plot these values on the same graph made in Step 2 above.
6. Read the coordinate points (q, P_{wf}) of the intersection between the two curves (IPR and TPR), see Figure 3.33. It is obtained that the maximum possible rate is 332 stb/d.

2. The productivity index J of two-phase fluids may be defined as:

$$J = -dq/dP_{wf} = -\left[-0.2\frac{q_{max}}{P_r} - 1.6\frac{q_{max}}{P_r}(P_{wf}) \right]$$

$$= 0.2 \times \frac{515}{(2400)} + 1.6 \times \frac{515}{2400^2}(1325)$$

$$= 0.232 \text{ stb/d/psi}$$

Example 3.8

Suppose we have more test data for Example 3.7. The additional data given are FE = 0.7. (1) What is the AOFP when the skin effect is removed (FE = 1.0)? (2) What is the actual maximum potential (AOFP) of the well? (3) Determine the maximum possible rate. (4) Determine the maximum possible rate if no damage occurred.

Solution

1. Using Equation 3.48, for V = 0.8,

$$\frac{500}{q_{max}^{FE=1}} = 0.7\left[1 + 0.8\left(\frac{250}{2400}\right)\right]\left[1 - \left(\frac{250}{2400}\right)\right]$$

$$q_{max}^{FE=1} = 736 \text{ stb/d}$$

2. The maximum potential will occur when $P_{wf} = 0$,

$$q_{max}^{FE=1} = q_{max}^{FE=1} \times 0.7\left[1 + 0.8\left(\frac{0}{2400}\right)\right]\left(1 - \frac{0}{2400}\right)$$

$$= 515.2 \text{ stb/d}$$

3. The maximum rate possible is 332 stb/d (already determined in Example 3.7).

4. Construct an IPR with FE = 1.0. Vogel's equation is valid now,

$$q_0 = 736\left[1 - 0.2\left(\frac{P_{wf}}{2400}\right) - 0.8\left(\frac{P_{wf}}{2400}\right)^2\right]$$

P_{wf}, psi	q, stb/d
500	680.0
800	621.5
1200	515.2
1500	414.0
2000	204.4

Plotting P_{wf} versus q, we can determine that $q = 477$ stb/d (Figure 3.33).

Example 3.9

An oil well is produced at $P_{wf} > P_b$. Data given are:

H = 5000 ft.
d = 2.375 in.
P_r = 1500 psi
J = 0.4 stb/d/psi
GLR = 0.8 Mscf/stb

Find the production rate of liquid and gas if (a) the bean size is $\frac{22}{64}$ in. or (b) the bean size is $\frac{30}{64}$ in. Assume critical flow conditions and use Gilbert's equation.

Solution

Since the well is operated above the reservoir bubble point, we can treat the inflow performance as a linear one:

$$P_{wf} = P_r - \frac{q}{J} = 1500 - \frac{2}{0.4}$$

Chose some values of q (e.g., 100, 200, and 400 stb/d) and use the gradient curves for 2.375-in. tubing. Gilbert's curves are used here for convenience. For a given P_{wf}, P_{wh} is found with a procedure opposite to that for determining P_{wf} for a known P_{wh} (solution to Example 3.7, (1), Step 4). Doing so, we can get the following:

q, stb/d	P_{wf}, psi	P_{wh}, psi
100	1250	540
200	1000	400
400	500	40

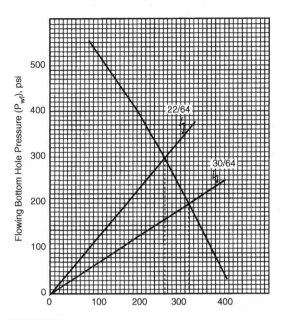

FIGURE 3.34   Pressure rate relationship for Example 3.9.

Plot q versus P_{wh} (shown in Figure 3.34). Generate other q versus P_{wh} for the two choke sizes using:

$$P_{wh} = 435 R^{0.546} q / S^{1.89}$$

Where (a) $q = (22^{1.89} \times P_{wh})/(435 \times 0.8^{0.546})$, or (b) $q = (30^{1.89} \times P_{wh})/(435 \times 0.8^{0.546})$

These are straight line equations through the origin. Draw the choke performance curves (see Figure 3.34). From this figure, we can determine that:

1. installing a $\frac{22}{64}$-in. choke:

$$q_{liquid} = 264 \text{ stb/d}$$
$$q_{gas} = 264 \times 0.8 \text{ M} = 0.211 \text{ MMscf/d}$$

2. installing a 30/64-in. choke:

$$q_{liquid} = 319 \text{ stb/d}$$
$$q_{gas} = 0.255 \text{ MMscf/d}$$

Example 3.10

Constructing an IPR curve of an oil well is sometimes difficult in terms of data available. The present reservoir pressure is not available or measured.

Some oil companies do not want to lose their production caused by closing the well to measure the static reservoir pressure. A practical means of overcoming this problem is to test the well at two different rates while measuring the flowing bottomhole pressure.

This technique does not require expensive, time-consuming tests. The pressure measurements can be very accurate, using a subsurface gage on flowing wells; they can be obtained with surface-recording downhole gage; or they can simply be obtained with casing pressure and fluid level shots, depending on the well condition.

A typical two-point test datum is taken from a well in the Judy Creek Beaverhill Lake 'A' Pool producing under a solution gas drive [24].

	Test 1	Test 2
q, stb/d	690.6	470
P_{wf}, psi	1152.0	1970
P_b, psi	2290.0	2290

The task is to estimate the static reservoir pressure and to construct IPR for this well.

Solution

$$\left(\frac{q}{q_{max}}\right)_1 = 1 - 0.2\left(\frac{1152}{2290}\right) - 0.8\left(\frac{1152}{2290}\right)^2 = 0.697$$

$$\left(\frac{q}{q_{max}}\right)_2 = 1 - 0.2\left(\frac{1970}{2290}\right) - 0.8\left(\frac{1970}{2290}\right)^2 = 0.236$$

$$q_c = (470 - 690.6)/(0.236 - 0.697) = 478.5 \text{ stb/d}$$

$$= (q_{max} - q_b)$$

Flowrate at bubble point,

$$q_b = 470 - (0.236)(478.5) = 357 \text{ stb/d}$$

$$J_b = 1.8 \times (478.5/2290) = 0.376$$

$$P_r = 2290 + (357/0.376) = 3239 \text{ psia}$$

The calculated reservoir pressure from the two-point test was 3239 psia, which is consistent with static pressures measured in this area of the 'A' Pool.

Figure 3.35 is the constructed IPR. A third test was run to verify this curve, with a rate of 589 stb/d at 1568 psia. This point fell essentially on the curve generated by the original two-test points.

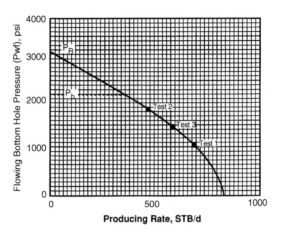

FIGURE 3.35   Pressure-rate relationship for Example 3.10.

Example 3.11

A reservoir with a back-pressure curve slope $(1/n) = 1.12$ has the following two flow tests:

	$\overline{P}$, psia	q_o, stb/d	P_{wh}, psia
Test 1	2355.9	335.1	1300
Test 2	2254.9	245.8	13

Calculation of future IPR:

- Using Fetkovich's method:

$$q_{max} = \frac{q_0}{\left[1 - \left(\dfrac{P_{wf}}{P}\right)^2\right]^n}$$

$$q_{max1} = \frac{335.1}{\left[1 - \left(\dfrac{1300}{2355.4}\right)^2\right]^{0.893}} = 463.5 \text{ stb/d}$$

$$q_{max2} = \frac{254.8}{\left[1 - \left(\dfrac{1300}{2254.9}\right)^2\right]^{0.893}} = 365.5 \text{ stb/d}$$

- Calculate J^*:

$$J_1^* = \frac{463.5}{2355.4} = 0.197 \text{ stb/d/psi}$$

$$J_2^* = \frac{365.5}{2254.9} = 0.162 \text{ stb/d/psi}$$

- Calculate A' and B':

$$A' = \frac{0.197 - 0.162}{(2355.4)^2 - (2254.9)^2} = 7.554 \times 10^{-8}$$

$$B' = \frac{\dfrac{0.197}{(2355.4)^2} - \dfrac{0.162}{(2254.9)^2}}{\dfrac{1}{(2355.4)^2} - \dfrac{1}{(2254.9)^2}} = -0.222$$

- Calculate maximum future rate (for instance, a future reservoir pressure of 1995 psia):

$$q_{max} = 7.554 \times 10^{-8}(1995)^3 + (-0.222)(1,995)$$
$$= 157 \text{ stb/d}$$

- The future IPR curve can then be predicted using the equation

$$q_0 = 157\left[1 - \left(\frac{P_{wf}}{1995}\right)^2\right]^{0.893}$$

References

[1] Jones, L. G., Blount, M., and Glaze, O. H., "Use of Short Term Multiple Rate Flow Tests to Predict Performance of Wells Having Turbulence," Paper SPE 6133, prepared for the 51st Annual Fall Technical Conference and Exhibition of the SPE of AIME, New Orleans, Louisiana, October 3–6, 1976.

[2] Crane Company Industrial Products Group, "Flow of Fluid Through Valves, Fittings and Pipe," Chicago, Illinois, Technical Paper no. 410.

[3] Rawlins, E. L., and Schellhardt, M. A., "Back-Pressure Data on Natural Gas Wells and Their Application to Production Practices," Monograph 7, USBM, 1936.

[4] Mishra, S., and Caudle, B. H., "A Simplified Procedure for Gas Deliverability Calculations Using Dimensionless IPR Curves," paper SPE 13231, presented at the 59th Annual Technical Conference and Exhibition, Houston, Texas, September 16–19, 1984.

[5] Dake, L. P., "*Fundamentals of Reservoir Engineering*," Elsevier Scientific Publishing Co., Amsterdam, 1978.

[6] Chase, R. W., and Anthony, T. M., "A Simplified Method for Determining Gas Well Deliverability," paper SPE 14507, presented at the SPE Eastern Regional Meeting, Morgantown, West Virginia, November 6–8, 1985.

[7] Cullender, M. H., "The Isochronal Performance Method of Determining Flow Characteristics of Gas Wells," *Transactions of AIME*, 1955.

[8] Cullender, M. H., and Smith, R. V., "Practical Solution of Gas-Flow Equations for Wells and Pipelines with Large Temperature Gradients," *Transactions of AIME*, 1956.

[9] Katz, D. L., et al., *Handbook of Natural Gas Engineering*, McGraw-Hill Book Co., New York, 1959.

[10] Peffer, J. W., Miller, M. A., and Hill, A. D., "An Improved Method for Calculating Bottom-hole Pressure in Flowing Gas Wells with Liquid Present," paper SPE 15655, presented at the 61st Annual Technical Conference and Exhibition of the SPE, New Orleans, Louisiana, October 5–8, 1986.

[11] Nind, T. E. W., *Principles of Oil Well Production*, 2nd Edition, McGraw-Hill Book Company, New York, 1981.

[12] Ikoku, C. U., *Natural Gas Production Engineering*, John Wiley & Sons, Inc., New York, 1984.

[13] Vogel, J. V., "Inflow Performance Relationships for Solution-Gas Drive Wells," *Journal of Petroleum Technology*, January 1968.

[14] Fetkovich, M. J., "The Isochronal Testing of Oil Wells," paper SPE 4529, prepared for the 48th Annual Fall Conference and Exhibition of the SPE of AIME, Las Vegas, Nevada, September 30–October 3, 1973.

[15] Standing, M. B., "Inflow Performance Relationships for Damaged Wells Producing by Solution-Gas Drives," *Journal of Petroleum Technology*, November 1970.

[16] Camacho, V., Raghavan, R. G., "Inflow Performance Relationships for Solution-Gas Drive Reservoirs," *Journal of Petroleum Technology*, May 1989.

[17] Standing, M. B., "Concerning the Calculation of Inflow Performance of Wells Producing from Solution-Gas Drive Reservoirs," *Journal of Petroleum Technology*, September 1971.

[18] Kelkar, B. G., and Cox, R., "Unified Relationship to Predict Future IPR Curves for Solution-Gas Drive Reservoirs," paper SPE 14239, prepared for the Annual Technology Conference and Exhibition of the SPE of AIME, Las Vegas, Nevada, September 22–25, 1985.

[19] Brown, K. E., *The Technology of Artificial Lift Methods*, Vols. 3a and 3b, Petroleum Publishing Co., Tulsa, Oklahoma, 1980.

[20] Gilbert, W. E., "Flowing Gas Well Performance," *Drilling and Production Practices*, 1954.

[21] Poettmann, F. E., and Beck, R. L., "New Charts Developed to Predict Gas-Liquid Flow Through Chokes," *World Oil*, March 1963.

[22] Ashford, F. E., "An Evaluation of Critical Multiphase Flow Performance Through Wellhead Chokes," *Journal of Petroleum Technology*, August 1974.

[23] Sachdeva, R., Schmidt, Z., Brill, J. P., and Blais, R. M., "Two-phase flow through chokes," paper SPE 15657, presented at the 61st Annual Technical Conference and Exhibition of the SPE, New Orleans, Louisiana, October 5–8, 1986.

[24] Richardson, J. M., and Shaw, A. H., "Two-Rate IPR testing—A practical production tool," *The Journal of Canadian Petroleum Technology*, March–April, 1982.

Gas Production Engineering

4.1 SURFACE PRODUCTION/SEPARATION FACILITY

The purpose of the surface production facility (Figures 4.1 and 4.2) is:

- To separate the wellstream into its three fundamental components—gases, liquids, and solid impurities
- To remove water from the liquid phase
- To treat crude oil to capture gas vapors
- To condition gas

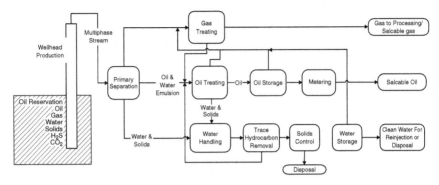

FIGURE 4.1   Typical oil production process system.

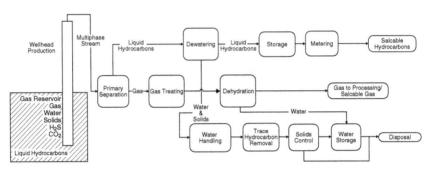

FIGURE 4.2   Typical gas production process system.

This section is a discussion of the design, use, function, operation, and maintenance of common facility types. In this section facilities will be referred to as the following:

- Oil production facilities as *batteries*
- Primary gas production facilities as *gas units*
- Water handling/disposal as *brine stations*

Identification of the service of each facility and associated equipment is also required. Is the service natural gas, crude oil, produced water, multiphase? Is it sweet (little to no H_2S) or sour (H_2S present)? H_2S or sour service is ambiguous at best without the following:

- National Association of Corrosion Engineers (NACE) MR-01-75 addresses material selection to avoid sulfide stress cracking. Threshold concentrations for application are in the Scope section of this document. In general, MR-01-75 applies to wet gas steams (where free water is present) when operating pressure exceeds 50 psig and H_2S is present, or to multiphase systems when pressures exceed 350 psig and

H_2S is present. The Minerals Management Service (MMS) cautions that even though materials listed may be resistant to sulfide stress corrosion environments, they may not be suitable for use in chloride stress cracking environments [1].

- Where H_2S concentration exceeds 100 ppm, The Texas Railroad Commission Rule 36 [2] and New Mexico OCD Rule 118 [3] apply. Both these rules address public safety. Other states have similar rules.
- Many oil and gas companies also have recommended standards for various levels of H_2S concentration, so be aware that these may also need to be taken into account.

Before anything is designed, construction codes need to be reviewed as well. Some of the most referenced codes are:

- American National Standards Institute (ANSI) B31.8 "Gas Transmission and Distribution Piping Systems" is used for general oil field construction. ANSI B31.8 dictates that the qualification of welders and the quality of fit-up, taking, welding, and radiographic inspection for noncritical service should conform to API Standard 1104 "Standard for Field Welding of Pipelines".
- ANSI B31.3 "Chemical Plant and Petroleum Piping" applies to all process piping in gas plants and compressor station installation with pressures over 750 psig.
- Title 49 of the Code of Federal Regulations (CFR) covers Department of Transportation (DOT) regulations in parts 191, 192, and 195 [4–6].
- As always, be aware that various other state and federal requirements may apply.

Figure 4.3 is an example process flow diagram of surface production equipment more commonly referred to as a *battery*.

From a technical point of view, production equipment can be divided into four major groups:

- pressure vessels
- storage tanks
- prime movers (pumps and compressors)
- piping

4.1.1 Nomenclature of Separating Pressure Vessels

Pressure vessels in the oilfield are used for separating well fluids produced from oil and gas wells into the gaseous and liquid components. Vessels are separated into two groups by the American Society of Metallurgical Engineering (ASME), the "governing" body of pressure vessel standards and codes [7]. These are low-pressure, less than or equal to 125 psig,

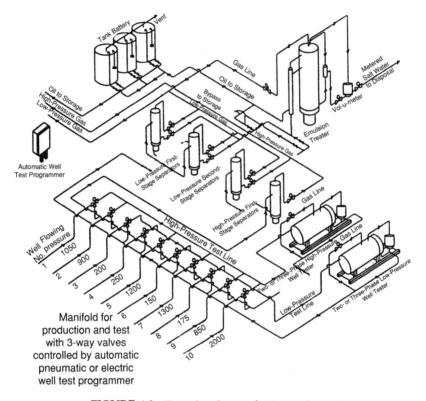

FIGURE 4.3   Typical surface production equipment.

or high-pressure, greater than 125 psig. Some of the more common types of vessels used for separation in oil and gas production are:

- *Separator*—a vessel used to separate a mixed-phase stream into gas and liquid phases that are "relatively" free of each other. Other terms used are scrubbers, knockouts, line drips, and descanters. The design of these vessels may be either low or high pressure.
- *Scrubber or Gas Knockout*—a vessel designed to handle streams of high gas/liquid ratios. The liquid is generally entrained as mist in the gas or is free, flowing along the pipe wall. These vessels usually have a small liquid collection section. The terms scrubber and knockout are used interchangeably. Typically, these are high-pressure vessels.
- *Three–Phase Separator*—a vessel used to separate gas and two immiscible liquids of different densities (e.g., gas, oil, and water). Typically, these vessels are low pressure but can just as easily be designed for high pressures.
- *Liquid-Liquid Separator or Free Water Knockout*—two immiscible liquid phases can be separated using the same principles as for gas and liquid separators. These vessels are designed to operate at much lower

velocities and pressures than the gas-liquid separators; they are designed to take advantage of the physical properties of the liquids, and inasmuch vertical vessels may be more effective than horizontal.

- *Filter Separator*—These separators usually have two compartments. The first contains filter/coalescing elements. As the gas flows through the elements, the liquid particles coalesce into larger droplets, and when the droplets reach sufficient size, the gas flow causes them to flow out of the filter elements into the center core. The particles are then carried into the second compartment of the vessel (containing a vane-type or knitted wire mesh mist extractor) where the larger droplets are removed. A lower barrel or boot may be used for surge or storage of the removed liquid.
- *Line Drip*—Typically used in pipelines with very high gas/liquid ratios to remove only free liquid from the gas stream and not necessarily all the liquid from a gas stream. Line drips provide a place for free liquids to separate and accumulate. Typically, these are high-pressure vessels.
- *Slug Catcher*—A separator, which is designed to absorb sustained inflow of large liquid volumes at irregular intervals. Usually found on gas-gathering systems or other two-phase pipeline systems. A slug catcher may be a single large vessel or a manifold system of pipes. Typically, these are high-pressure vessels.
- *Flash Tank (Chamber, Trap, or Vessel)*—A conventional oil and gas separator operated at low pressure, with the liquid from a higher-pressure separator being "flashed" into it. This flash chamber is quite often the second or third stage of separation, with the liquid being discharged from the flash chamber to storage.
- *Expansion Vessel*—A first-stage separator on a low-temperature or cold-separation unit. This vessel may be equipped with a heating core to melt hydrates, or a hydrate-preventative liquid (such as glycol) may be injected into the well fluid just before expansion into this vessel.
- *Desanders*—A high-pressure vessel that is installed on some gas wells to "catch" any sand that is produced along with the gas. This sand may be from a frac job, or it may be from the formation. The pressure drop at the desander allows the majority of this sand to drop out at a controlled location rather than in the lines, at measurement points, or in production equipment.

All new pressure vessels should be constructed per ASME code. It is up to the purchaser if the vessel is to be certified and registered by the National Board of Boiler and Pressure Vessel Inspectors [8]; in critical service, this may be worth the extra time and expense to obtain a National Board Number and the associated paperwork. It is imperative that all the construction drawings, steel mill reports, and the U1A form that is provided by the ASME Code Shop/Fabricator be filed in a Central Filing System. Those records are required to be maintained and accessible as part of the Code requirements. Any subsequent repair or alterations made to the vessel are to be documented and filed as well. It is suggested that a unique company number be assigned to each vessel and cold stamped into a vessel leg or skirt, as nameplates can be damaged or removed over time, especially if equipment is moved from one location to another. This unique number allows for maintaining the records from cradle to grave for each vessel.

Utilizing used equipment is a matter of timing and economics. The cost of new versus used needs to be taken into account on all projects. The cost of used equipment needs to include inspection, repair, and certification costs as well as the cost of the vessel itself.

All equipment should be upgraded and tested to a level that meets regulations and industry standards (i.e., API) for operability and safety. Used pressure vessels should conform to API 510 standards [9].

4.2 PRESSURE VESSEL RELIEF SYSTEMS

All pressure vessels are to be equipped with a pressure relief valve set to open at a pressure no greater than the maximum allowable working pressure (MAWP) of that vessel and sized to prevent pressure from rising more than 10% or 3 psi, whichever is greater, above the MAWP at full open flow. For vessels with multiple relief valves, 16% or 4 psi, whichever is greater, above the MAWP is allowed. This provision **DOES NOT** apply to relief valve/rupture disk combinations. Pressure vessels operating at pressures above 125 psig may be equipped with a rupture disk or a second (redundant) relief valve. The rupture disk should be designed to rupture at not more than 15% above the MAWP and sized to handle the maximum flow through the vessel. Relief valves and rupture disks should be installed in the gaseous phase of a vessel, and be placed so that there is no restriction between the valve and the vessel. A locking handle full-opening ball valve can be installed below the relief valve but must remain locked in the open position. The installation of this isolation valve is so that the relief valve can be tested in place or replaced without having to take the vessel out of service to do so. Vessels that operate in a fluid-packed state should be equipped with an appropriately sized liquid service relief valve.

4.2.1 Product Storage

Tanks used in oil and gas production operations are all atmospheric. Gas plants and refineries use the pressurized tanks/vessels.

- *Atmospheric*—Atmospheric tanks are designed and equipped for storage of contents at atmospheric pressure. This category usually employs tanks of vertical cylindrical configuration that range in size from small

TABLE 4.1 API standards—tanks (partial listing) [10]

Standard Number	Title
API Spec 12B	Specification for Bolted Tanks for Storage and Production Liquids
API Spec 12D	Specification for Field Welded Tanks for Storage and Production Liquids
API Spec 12F	Specification for Shop Welded Tanks for Storage and Production Liquids
API Spec 12P	Specification for Fiberglass Reinforced Plastic Tanks
API 12R 1	Recommended Practice: Setting, Maintenance, Inspection, Operation and Repair in Production Service
API RP 620	Recommended Practice: Design and Construction of Large, Welded, Low-Pressure Storage Tanks
API RP 650	Recommended Practice: Welded Steel Tanks for Oil Storage
API Spec 653	Specification for Tank Inspection, Repair, Alteration and Reconstruction
API RP 2000	Recommended Practice: Venting Atmospheric and Low-Pressure Storage Tanks: Non-Refrigerated and Refrigerated

shop welded to large field erected tanks. Bolted tanks, fiberglass tanks, and occasionally rectangular welded tanks are also used for atmospheric storage. API standards relating to tankage are listed in Table 4.1 [10]. Note: 1,000 bbl tanks are the largest size that can be shop welded and transported to location. Internally coated, welded steel tanks should normally be used in hydrocarbon or brine service. Tank decks should be connected to ensure proper internal coating. Operations preferences and service conditions should be considered when selecting flat or cone bottom tank. Fiberglass tanks are typically used in brine water service rather than for hydrocarbon due to fire safety. It is important to always ground the fluid in a fiberglass tank, bond and ground metal parts, and gas blanket the vapor space to minimize oxygen entrainment.

- Low Pressure (0 to 2.5 psig or 0 to 17 kPa)—Low-pressure tanks are normally used for storage of intermediates and products that require a low internal gas pressure. These are generally cylindrical in shape with flat or dished bottoms and sloped or domed roofs. Low-pressure storage tanks are usually of welded design; however, bolted tanks are often used for operating pressures near atmospheric. Many refrigerated storage tanks operate around 0.5 psig. These tanks are built to API Standards 12D, 12F, 620, or 650.
- Medium Pressure (2.5 to 15 psig or 17 to 103 kPa)—Medium-pressure tanks are normally used for storage of higher volatility intermediates and

products. The shape may be cylindrical with flat or dished bottoms and sloped or domed roofs. Medium-pressure tanks are usually of welded design. Welded spheres may also be used, particularly for pressure at or near 15 psig. These tanks are built to API Standard 620.

- High Pressure (above 15 psig or 103 kPa)—High-pressure tanks/ vessels are generally used for the storage of refined products or fractionated components at pressures greater than 15 psig. They are welded and may be of cylindrical or spherical configuration. Because they are above 15 psig, they are designed and constructed per ASME code.

Gas processing industry liquids are commonly stored underground, in conventional or solution-mined caverns. For more details about product storage options, see Table 4.2.

4.2.1.1 Walkways and stairways

Engineering controls should be installed on new tanks and walkways so that breathing equipment is not required to do routine work. Engineering controls include mechanical or electronic tank gauges that eliminate the need for daily opening of the thief hatch to measure the tank level. All readouts should be at ground level. Walkways and stairways should be constructed to API standards and be wide enough so that a person with a self-contained breathing apparatus (SCBA) can have egress without restriction

TABLE 4.2   Product Storage Options [11]

	Atmospheric	Low Pressure (0–2.5 psig)	Medium Pressure (2.5–15 psig)	High Pressure (Above 15 psig)	Underground
Crude oil	X	X	X	–	X
Condensate	X	X	X	X	X
Refined oil	X	X	–	–	X
Gasoline	X	X	X	–	X
Butane	–	X_R	X_R	X	X
Propane	–	X_R	X_R	X	X
Raw NGLs	–	X_R	X_R	X	X
Ethane	–	X_R	X_R	X	X
Petrochemicals	–	X_R	X_R	X	X
Natural gas	–	–	–	X	X
LNG	–	X_R	X_R	X	–
Treating agents	X	X	–	–	–
Dehydration liquids	X	X	–	–	–
Chemicals	X	X	X	–	–
Solids	X	–	–	–	–
Produced water	X	–	–	–	–

X_R, refrigerated only.

(minimum of 30 in. [0.8 m] wide). Walkways should be quoted complete with all necessary supports. Engineers should carefully review standard walkways and stairways provided by manufacturers, as some do not meet the structural requirements of API or OSHA.

4.2.1.2 Tank Venting

All tank venting designs should be in accordance to API 2000 Standard. Oil and gas production operation tanks should be provided with both normal and emergency venting. Normal venting is typically addressed with deadweight pressure/vacuum valves. The weight is determined by the design of the tank. Tanks used for production operations typically have vent valves set at 4 oz. Emergency venting is addressed with gauge hatch covers and, if necessary, additional weighted covers. The normal operation vents must work in conjunction with the gas blanket and the vapor recovery unit (VRU) (more information on VRUs is found in "Compressors"), if installed. Frangible deck joints are recommended on larger tanks. These should be constructed according to API 650 with full penetration welds to facilitate internal coating of the tank.

4.2.1.3 Gas Blankets

Gas blankets should be used in order to minimize the possibility of an explosive atmosphere in water tanks, to prevent oxygenation of produced water, and to prevent air at the intake of a VRU if the controls fail. Gas blankets must work in coordination with the VRU and normal tank venting. Normal relief pressures should be at the widest range permissible by tank design. The source of gas for a gas blanket is typically a residue gas system, which also provides "clean" gas for fired equipment or may come off of the gas scrubber in the facility.

4.2.2 Prime Movers

4.2.2.1 Centrifugal Pumps

Centrifugal pumps are the most commonly used pumps in production operations. Single-stage ANSI pumps driven by small-horsepower electric motors are used for transfer of fluids as well as charge pumps upstream of larger high-pressure water injection pumps. ANSI pumps are preferred in production operations because all similar pumps meeting ANSI specifications will have identical baseplate, as suction/discharge dimensions, regardless of the manufacturer. In addition, this allows for greater future utility and ease of maintenance and replacement.

Materials should be specified according to service and design life requirements. Carbon steel housing may be suitable for sweet oil service, but in corrosive service, alloys such as 316 stainless steel (SS) are justified. When sizing a pump, it is necessary to take into account current and future operating conditions in order to provide for flexibility after installation. It is much easier and is more economic to install a larger impeller in a pump than it is to replace the motor and/or pump. The engineer needs to find a balance between conservative pump size and being able to keep the pump operating at or near its best efficiency. As operating conditions fall too far to one side or the other, surge effects and cavitation can occur.

Multistage centrifugal pumps are commonly used for water injection. These pumps can move large volumes of water at relatively high pressure (less than 2,000 psi). Because of the corrosive service, these pumps are typically made of 316 SS. They may be horizontal spilt case or vertical "can" design. If at all possible, it is suggested that there be a spare pump set up to switch to if problems arise with the primary pump.

4.2.2.2 Reciprocating Positive Displacement Pumps

In a reciprocating pump, a volume of liquid is drawn into the cylinder through the suction valve on the intake stroke and is discharged under positive pressure through the outlet valves on the discharge stroke. The discharge from a reciprocating pump is pulsating and changes only when the speed of the pump is changed. This is because the intake is always a constant volume. Often an air chamber is connected on the discharge side of the pump to provide a more even flow by evening out the pressure surges. Reciprocating positive displacement (PD) pumps are most often used for water injection or disposal. They are designed for high pressures and low volumes as compared to the multistage centrifugal pumps. Typically, the fluid end is 316 SS due to the corrosive nature of the fluids handled. As always, know the service conditions before specifying the material. When sizing a PD pump, it is suggested that the initial (design) speed be less than 90% of the manufacturer's rated speed. This will enhance the reliability of the machine and allows some flexibility for increasing flow rate if necessary. By changing the speed and the plunger diameters, a fluid end can address a range of flow rates and pressures.

Reciprocating PD pumps are available in four designs:

- *Simplex*—This pump has a single liquid cylinder that forces liquid out through the top outlet on both the in and the out stroke (here up and down). This basic type of pump might be used for air pumps, feed pumps for the furnace, fire, bilge, and fuel oil service.
- *Duplex*—These are similar to the simplex pump, having two pistons instead of one, providing smoother operation. Duplex pumps have no "dead spots" because one or the other steam piston is always under force of steam (or compressed air). The two pistons are about 1/4 cycle out of synchronization with each other.
- *Triplex*—These are similar to the duplex pumps, having three pistons or plungers. This is the most commonly used pump in drilling and well

servicing. They are also widely used for disposal wells or small water-floods.

- *Quintuplex*—These are similar to the triplex pumps, having five pistons or plungers.

Piping design should carefully address the suction-side pipe hydraulics. It is critical to provide adequate Net Positive Suction Head (NPSH) to avoid cavitation and vibration problems. The use of suction and discharge pulsation dampeners is strongly recommended. Suction lines should be sized to limit the velocity to less than 5 ft./s, and discharge piping should include a relief valve and check valve upstream of the discharge block valve. The relief valve should be piped back to a tank to avoid spills.

Vibration is of major concern with PD pumps; mechanical vibration and structural stress can result in failures and spills, and so pulsation dampeners, valves, etc. should be supported. OSHA requires guards to be placed over plungers and belts.

4.2.2.3 Compressors

Compression in the oil field ranges from low-pressure sliding vane compressor to 2500+ high-pressure integral units. Compressors are used to boost the pressure from the wellhead or the facility to sales line pressure, to inject gas back into the reservoir for pressure maintenance or enhanced oil recovery (CO_2, nitrogen, etc.), or boost the pressure from the wellhead or facility to a central facility. Depending upon the size and the field conditions/location, they may be powered by electric motors, gas/internal combustion or by turbine engines. VRUs are the most common use of compression in surface oil production. The larger compressors are addressed in "Compressor Stations".

VRUs are installed for the economic recovery of tank vapors, compliance with Air Emission Regulations (Federal and/or State), and for safety considerations. Emissions can be estimated or measured prior to installation. The Texas Natural Resource Commission has allowed 90 days of temporary operation under Standard Exemption 67 [12] to determine actual emissions. The majority of the VRUs on the market today use sliding vane rotary compressors. In the low-pressure applications, a blower-type compressor should also be considered.

4.2.3 Piping Guidelines

4.2.3.1 Materials

The most commonly used materials in oil and gas operations are carbon steel, SS (304 or 316 families), fiberglass, and polyethylene. Each one has a place when properly matched to the application. In most cases, the decision on which one to use is based on economics (material costs as well as the coating and ditching costs), corrosion resistance, land use, support requirements, proximity to the public, and impact of failures. Much of what is addressed is risk-based analysis.

- *Carbon steel* is the most commonly used material. Most onshore operations use electric fusion welded (EFW), electric resistance welded (ERW), or grade B seamless pipe. The EFW and ERW pipe are considered acceptable for most applications and are cheaper than seamless pipe. EFW is typically only available in 16-in. or greater diameter. DOT-regulated lines cannot contain ERW pipe.

    Bare carbon steel can be used for liquid hydrocarbon, steam, and gas service. If internally coated, it can also be used in water and acid service.

- *SS* is used when increased corrosion protection is needed. The most commonly used ones in the oil field are 316 and 304. The low-carbon versions should be used if it is to be welded. The 316L and 304L do not become sensitized during welding; hence, do not require postweld heat-treating. 316 SS is preferred in brine/salt water service, and 304 SS is commonly used in wet CO_2 service.

- *Fiberglass* is commonly used where corrosion, either internal or external, is of concern. It is available in a wide variety of diameters and pressure ratings. It is primarily used in water systems (water injection and gathering systems as well as inplant piping at injection stations). Because of the effect of sunlight (UV) on the resins used in fiberglass pipe, it should be buried, painted, or otherwise protected from the elements. If buried, thrust blocks should be considered at the elbows to counteract the hydraulic forces.

- *Polyethylene* is commonly used due to its low cost, low weight, ease of installation, corrosion resistance, and flow characteristics. For the most part, it is not recommended for pressure above 100 psig or temperatures above 100°F, although there has been a lot of progress made in the characteristics. Poly pipe is most often used for gas gathering and flowlines, but it should be derated per manufacturer's recommendations when used in wet hydrocarbon service. Because of its composition, it should not be used to transport flammable fluids above ground.

4.2.3.2 Buried versus Above Ground

One of the first of these items to be addressed is whether the pipe is to be above ground or buried. OSHA requires that all nonmetallic or low-melting-point pipe carrying flammable materials to be buried. And whenever nonmetallic pipe is buried, detection tape or trace cable should be buried above the pipe to aid in locating the pipe in the future. Buried pipe may be more or less costly than aboveground systems. One should take into account the relative cost of the pipe, access, operational preferences, fire, freeze, mechanical protection, and pipe supports. It is recommended to externally coat buried pipe to avoid external corrosion. Cathodic protection may be used instead of or in conjunction with external coating systems. It is common to paint aboveground pipe for aesthetics as well as to provide some corrosion protection (this includes painting fiberglass pipe to protect it from UV degradation). Aboveground piping should be labeled for the type of service and carrier fluid. ASME A13.1 "Scheme for the Identification of Piping Systems" can be referred to where color-coding and method of identification is required.

4.2.3.3 Internal Coatings and Linings

Solids production and erosion can be detrimental to some internal coatings, so these as well as the specific service and considerations of environmental risk associated with a leak should be evaluated. In most cases, internal coating should be considered for all steel piping systems. There are many different products available; the ones most commonly found in the oil field are cement lining, plastic coating, and extruded polyethylene.

4.2.3.4 Piping Connections

The following parameters need to be considered when determining what connections are best suited for a project:

- operations preference and experience (You *have* to get "buy-in" from the field personnel)
- operating pressure of the systems
- hydrocarbon content in the produced fluid
- H_2S concentrations in the product stream
- impact on surrounding area in the event of a failure
- expected life of the project

4.2.3.5 Flanged versus Threaded

Welded/flanged construction may provide the following advantages:

- less susceptible to mechanical damage (especially in vibrating service)
- greater structural joint strength
- more resistant to fatigue failure
- better hydrocarbon containment in the event of fire
- lower fugitive emissions
- make-ups are more precise than threaded

Threaded construction may provide the following advantages:

- less expensive to install
- easier and less expensive to repair
- no hot work involved
- quicker to disassemble

4.2.3.6 Valves

Types of valves commonly used in oil and gas production

- *Gate Valves*—A gate is a linear-motion valve that uses a typically flat closure element perpendicular to the process flow, which slides into the flow stream to provide shutoff. With a gate valve, the direction of fluid flow does not change, and the diameter through which the process fluid passes is essentially equal to that of the pipe. Hence, gate valves tend to have minimal pressure drop when opened fully. Gate valves are designed to minimize pressure drop across the valve in the fully opened position and stop the flow of fluid completely. In general, gate valves are not used to regulate fluid flow. A gate valve is closed when a tapered disk of a diameter slightly larger than that of the pipe is lowered into position against the valve seats. The valve is fully open when the disk is pulled completely out of the path of the process fluid into the neck [13]. Typically, these are used on pipelines, on gas wells, and as plant block valves.
- *Globe Valves*—A globe valve is a linear-motion valve characterized by a body with a longer face-to-face that accommodates flow passages sufficiently long enough to ensure smooth flow through the valve without any sharp turns. Globe valves are widely used to regulate fluid flow in both on/off and throttling service. The amount of flow restriction observed with valve disk (or globe) location is relative to the valve seat. The valve seat and stem are rotated 90° to the pipe. The direction of fluid flow through the valve changes several times, which increases the pressure drop across the valve. In most cases, globe valves are installed with the stem vertical and the higher-pressure fluid stream connected to the pipe side above the disk, which helps to maintain a tight seal when the valve is fully closed. Traditionally the valve disk and seat were both metal, although some modern designs use an elastomer disk seal. These valves are inexpensive and simple to repair [13].
- *Plug Valves*—Plug valves are similar to gates and work particularly well in abrasive service. These are available either lubricated or nonlubricated.
- *Ball Valves*—Ball valves are quarter-turn, straight-through flow valves that use a round closure element with matching rounded elastometric seats that permit uniform sealing stress. These valves are limited to moderate temperature service (below 250°C) by the plastic seats that create a seal around the ball. The type of seat can vary with the valve pressure rating and materials of construction. Some valve seats are a single molded form, while other valve seats with higher-pressure ratings often incorporate a Trunion design where each face of the ball is sealed separately. They have found applications in flow control, as well as on/off use in isolating a pipe stream. The pressure drop across the

valve in a fully open position is minimal for a full-port design. However, with restricted-port designs the pressure drop can be significant [13].

- *Butterfly Valves*—The butterfly valve is a quarter-turn rotary-motion valve that uses a round disk as the closure element. The sealing action of a butterfly valve is achieved by rotating a disk of approximately the same diameter, as the pipe from a globe valve is a function of the position in line with fluid flow to a position perpendicular to flow. The axial length of these valves is less than any other valve, which in cases in which flange faces are used with large pipe sizes (greater than 10 in.), these valves have the lowest initial cost. If resilient seats or piston rings on the disk are used, these valves can be sealed by relatively low operating torque on the valve stem. This sealing action is assisted by the fluid-pressure distribution that tends to close the valve. This same hydraulic unbalance requires that a latching device or worm gear be installed to prevent unwanted closure of manually operated valves. Although butterfly valves are used for low-pressure drop applications, the pressure drop across the valve is quite high with large flow rates compared with a gate valve [13]. Lug-type body is preferred in hydrocarbon applications because there is less bolt length exposed in a fire. The lug type is also easier to flange up than the wafer type because uneven tightening of the bolts in wafer type can cause uneven pull up and leaking. High-performance butterfly valves are commonly used for throttling in both low and high-pressure applications. A SS stem with a SS or a plastic coated disc is suggested for produced water service.
- *Check Valves*—Check valves commonly used in production service are the swing, piston, or split disk styles. The swing check is used in low-velocity flows where flow reversals are infrequent. Piston checks with or without spring assists are used in high–flow rate streams with frequent reversals (i.e., compressor or high-pressure pump discharges). Split disk checks are typically used in low-pressure applications in which a minimum pressure drop across the valve is required (i.e., transfer/charge pump discharges). Because of the service, SS internals are preferred over plastic-coated in any service where they come in contact with produced water to minimize corrosion failures.
- *Relief Valve*—A pressure relief valve is a self-operating valve that is installed in a process system to protect against overpressurization of the system. Relief valves are designed to continuously modulate fluid flow to keep pressure from exceeding a preset value. There are a wide variety of designs, but most resemble diaphragm valves, globe valves, or swing check valves. With many of these designs, a helical spring or hydraulic pressure is used to maintain a constant force that acts on the backside of the valve disk or diaphragm causing the valve to normally be closed. When the force exerted by the process stream (i.e., fluid pressure) on the valve disk is greater than the constant force exerted by the spring,

the valve opens, allowing process fluid to exit the valve until the fluid pressure falls below the preset value. These valves can be preset to a specific relief pressure, or they may be adjustable [13].

Materials used in valve construction are as varied as their applications. The following are the ones commonly used in the oil field. The material used in the valve body is addressed first. Valves carrying flammable fluids should be of steel or ductile iron.

Malleable iron should not be used in sour service because it does not meet NACE standards. Brass and bronze are suitable for fresh water and low-pressure air service. Nickel aluminum bronze and 316 SS are most often used in produced water applications. 304 SS is preferred for wet CO_2, whereas regular steel is acceptable for dry CO_2. Aluminum is acceptable for tank vent valves.

4.2.3.7 Valve Trim

Elastomer selection is critical. The most commonly used elastomer in brine water and hydrocarbon service is Buna-N (nitrile). The limitation is that it should not be used when temperatures exceed 200°F for prolonged periods. A 90-derometer peroxide-cured Buna is the first choice for use in CO_2 or H_2S service. Viton is good to about 350°F and is resistant to aromatics, but should not be used with amines or high pressure CO_2. Teflon is good to somewhere between 250°F to 350°F, depending on the application, and has exceptionally good chemical resistance characteristics. Teflon is not a true elastomer because it has no physical memory.

4.2.4   Pressure Vessel Design—Phase Separation [11, 14–16]

Practical separation techniques for liquid particles in gases are discussed. The principles used to achieve physical separation of gas and liquids or solids are momentum, gravity settling and coalescing. Any separator may employ one or more of those principles, but the fluid phases must be immiscible and have different densities for separation to occur.

4.2.4.1 Momentum

Fluid phases with different densities will have different momentum. If a two-phase stream changes direction sharply, greater momentum will not allow the particles of the heavier phase to turn as rapidly as the lighter fluid, so separation occurs. Momentum is usually employed for bulk separation of the two phases in a stream.

4.2.4.2 Gravity Settling

Liquid droplets will settle out of a gas phase if the gravitational forces acting on the droplet are greater than the drag force of the gas flowing

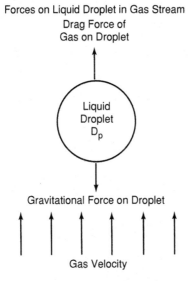

Forces on Liquid Droplet in Gas Stream
Drag Force of
Gas on Droplet

Liquid
Droplet
D_p

Gravitational Force on Droplet

Gas Velocity

FIGURE 4.4   Forces on liquid input in gas system [11].

around the droplet (Figure 4.4). These forces can be described mathematically. Drag force (F) on a liquid droplet in a gas stream is determine from

$$F = C_D A_p \gamma_g \frac{V_t^2}{2g} \tag{4.1}$$

Hence,

$$V_t = \left(\frac{2Fg}{C_D A_p \gamma_g} \right)^{0.5}$$

and from Newton's second law

$$F = \frac{w}{g} g \quad \text{and} \quad \text{if} \quad w = \frac{\gamma_L - \gamma_g}{\gamma_L} W_p$$

$$V_t = \left[\frac{2g W_p (\gamma_t - \gamma_g)}{\gamma_t \gamma_g A_p C_D} \right]^{0.5} = \left[\frac{4g D_P (\gamma_t - \gamma_g)}{3 \gamma_g C_D} \right]^{0.5} \tag{4.2}$$

where F = drag force in lb

V_t = terminal velocity in ft./s

g = acceleration due to gravity in ft./s^2

W_P = weight of particle in lb

$\gamma_{L,g}$ = liquid and gas-phase specific weights in lb/ft.^{3}

A_P = particle cross-sectional area in ft.^{2}

D_p = droplet diameter in ft.

C_D = drag coefficient of particle (dimensionless)

The drag coefficient has been found to be a function of the shape of the particle and the Reynolds number of the flowing gas. For the purpose of this equation, particle shape is considered to be a solid, rigid sphere.

Reynolds number is defined as

$$Re = \frac{1{,}488 D_p V_t \gamma_g}{\mu} \tag{4.3}$$

where μ = viscosity (cp).

In this form, a trial-and-error solution is required since both particle size D_p and terminal velocity V_t are involved. To avoid trial and error, values of the drag coefficient are presented in Figure 4.5 as a function of the product of drag coefficient C_D times the Reynolds number squared; this technique eliminates velocity from expression.

The abscissa of Figure 4.5 is represented by

$$C_D(Re)^2 = \frac{(0.95)(10^8)\gamma_g D_p^3 (\gamma_t - \gamma_g)}{\mu^2} \tag{4.4}$$

For production facility design (turbulent flow), the following formula for drag coefficient is proper:

$$C_D = \frac{24}{Re} + \frac{3}{(Re)^{0.5}} + 0.34 \tag{4.5}$$

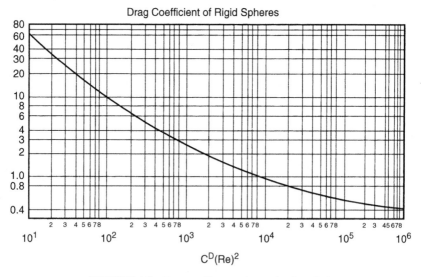

FIGURE 4.5   Drag coefficient of a rigid sphere [11].

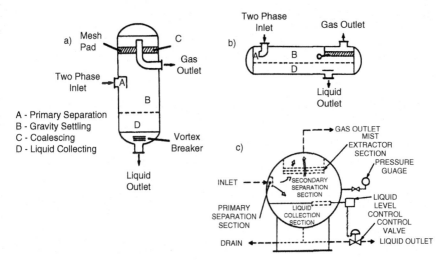

FIGURE 4.6 Gas liquid separators: (a) vertical, (b) horizontal, (c) spherical.

and if D_p is expressed in micrometers $\rightarrow d_m$

$$V_t = 0.0119 \left[\left(\frac{(\gamma_L - \gamma_g)}{\gamma_g} \right) \frac{d_m}{C_D} \right]^{0.5} \tag{4.6}$$

Equations 4.5 and 4.6 can be solved by an iterative solution as follows

1. Write the equation for laminar flows ($C_D = 0.34$)

$$V_t = 0.0204 \left[\frac{(\gamma_L - \gamma_g)}{\gamma_g} d_m \right]^{0.5}$$

2. Calculate $Re = 0.0049(\gamma_g d_m V_t / \mu)$
3. From Equation 4.5 calculate C_D
4. Recalculate V_t using Equation 4.6. Go to step 2

The above technique is proper assuming that known diameter drops are removed (e.g., 100 μm).

4.3 SEPARATOR DESIGN AND CONSTRUCTION

There are three types of separators: vertical, horizontal (single and double tube), and sometimes spherical (Tables 4.3 to 4.5).

TABLE 4.3   Standard Vertical Separators: Size and Working Pressure Ratings

Size (Dia. × Ht.)	Working Pressure (psi)								
16″ × 5′			230	500	600	1000	1200	1440	1500	2000
16″ × 7 1/2′			230	500	600	1000	1200	1440	1500	2000
16″ × 10′			230	500	600	1000	1200	1440	1500	2000
20″ × 5′			230	500	600	1000	1200	1440	1500	2000
20″ × 7 1/2′			230	500	600	1000	1200	1440	1500	2000
20″ × 10′			230	500	600	1000	1200	1440	1500	2000
24″ × 5′		125	230	500	600	1000	1200	1440	1500	2000
24″ × 7 1/2′		125	230	500	600	1000	1200	1440	1500	2000
24″ × 10′			230	500	600	1000	1200	1440	1500	2000
30″ × 5′			230	500	600	1000	1200	1440	1500	2000
30″ × 7 1/2′			230	500	600	1000	1200	1440	1500	2000
30″ × 10′		125	230	500	600	1000	1200	1440	1500	2000
36″ × 5′		125								
36″ × 7 1/2′		125	230	500	600	1000	1200	1440	1500	2000
36″ × 10′		125	230	500	600	1000	1200	1440	1500	2000
36″ × 15′			230	500	600	1000	1200	1440	1500	2000
42″ × 7 1/2′			230	500	600	1000	1200	1440	1500	2000
42″ × 10′			230	500	600	1000	1200	1440	1500	2000
42″ × 15′			230	500	600	1000	1200	1440	1500	2000
48″ × 7 1/2′			230	500	600	1000	1200	1440	1500	2000
48″ × 10′		125	230	500	600	1000	1200	1440	1500	2000
48″ × 15′		125	230	500	600	1000	1200	1440	1500	2000
54″ × 7 1/2′			230	500	600	1000	1200	1440	1500	2000
54″ × 10′			230	500	600	1000	1200	1440	1500	2000
54″ × 15′			230	500	600	1000	1200	1440	1500	2000
60″ × 7 1/2′			230	500	600	1000	1200	1440	1500	2000
60″ × 10′		125	230	500	600	1000	1200	1440	1500	2000
60″ × 15′		125	230	500	600	1000	1200	1440	1500	2000
60″ × 20′		125	230	500	600	1000	1200	1440	1500	2000

TABLE 4.4   Standard Horizontal Separators: Size and Working Pressure Ratings

Size (Dia. × Ht.)	Working Pressure (psi)								
12 3/4″ × 5′		230	500	600	1000	1200	1440	1500	2000
12 3/4″ × 7 1/2′		230	500	600	1000	1200	1440	1500	2000
12 3/4″ × 10′		230	500	600	1000	1200	1440	1500	2000
16″ × 5′		230	500	600	1000	1200	1440	1500	2000
16″ × 7 1/2′		230	500	600	1000	1200	1440	1500	2000
16″ × 10′		230	500	600	1000	1200	1440	1500	2000
20″ × 5′		230	500	600	1000	1200	1440	1500	2000
20″ × 7 1/2′		230	500	600	1000	1200	1440	1500	2000
20″ × 10′		230	500	600	1000	1200	1440	1500	2000
24″ × 5′	125	230	500	600	1000	1200	1440	1500	2000
24″ × 7 1/2′	125	230	500	600	1000	1200	1440	1500	2000
24″ × 10′	125	230	500	600	1000	1200	1440	1500	2000
24″ × 15′		230	500	600	1000	1200	1440	1500	2000

(Continued)

TABLE 4.4 (Continued)

Size (Dia. x Ht.)	Working Pressure (psi)								
30″ × 5′	125	230	500	600	1000	1200	1440	1500	2000
30″ × 7 1/2′	125	230	500	600	1000	1200	1440	1500	2000
30″ × 10′	125	230	500	600	1000	1200	1440	1500	2000
30″ × 15′		230	500	600	1000	1200	1440	1500	2000
36″ × 7 1/2′		230	500	600	1000	1200	1440	1500	2000
36″ × 10′	125	230	500	600	1000	1200	1440	1500	2000
36″ × 15′	125	230	500	600	1000	1200	1440	1500	2000
36″ × 20′		230	500	600	1000	1200	1440	1500	2000
42″ × 7 1/2′		230	500	600	1000	1200	1440	1500	2000
42″ × 10′		230	500	600	1000	1200	1440	1500	2000
42″ × 15′		230	500	600	1000	1200	1440	1500	2000
42″ × 20′		230	500	600	1000	1200	1440	1500	2000
48″ × 7 1/2′		230	500	600	1000	1200	1440	1500	2000
48″ × 10′	125	230	500	600	1000	1200	1440	1500	2000
48″ × 15′	125	230	500	600	1000	1200	1440	1500	2000
48″ × 20′		230	500	600	1000	1200	1440	1500	2000
54″ × 7 1/2′		230	500	600	1000	1200	1440	1500	2000
54″ × 10′		230	500	600	1000	1200	1440	1500	2000
54″ × 15′		230	500	600	1000	1200	1440	1500	2000
54″ × 20′		230	500	600	1000	1200	1440	1500	2000
60″ × 7 1/2′		230	500	600	1000	1200	1440	1500	2000
60″ × 10′	125	230	500	600	1000	1200	1440	1500	2000
60″ × 15′	125	230	500	600	1000	1200	1440	1500	2000
60″ × 20′	125	230	500	600	1000	1200	1440	1500	2000

TABLE 4.5 Standard Spherical Separators: Size and Working Pressure Ratings

Size (O.D.)	Working Pressures (psi)									
24″		230	500	600	1000	1200	1440	1500	2000	3000
30″		230	500	600	1000	1200	1440	1500	2000	3000
36″		230	500	600	1000	1200	1440	1500	2000	3000
41″	125									
42″		230	500	600	1000	1200	1440	1500	2000	3000
46″	125									
48″		230	500	600	1000	1200	1440	1500	2000	3000
54″	125									
60″		230	500	600	1000	1200	1440	1500	2000	3000

There are many vessel design software packages out on the market today, and everyone has a favorite. When the requirements are input and the program is run, one is then armed with the information needed to talk to the vendors about a bid specification or to look at what is available within the company's used equipment inventory. When purchasing a new vessel the data and requirements should always be provided to the supplier and reviewed by engineering on both sides to obtain the best vessel fit for each application.

Vertical separators are usually selected when the gas/liquid ratio is high or total gas volumes are low. In this sort of vessel, the fluids enter the vessel striking a diverting plate that initiates primary separation. Liquid removed by the inlet diverter falls to the bottom of the vessel. The gas moves upward, usually passing through a mist extractor to remove suspended mist, and then flows out. Liquid is removed by reservoir in the bottom. Mist extractors can significantly reduce the required diameter of vertical separators. However, if the production stream is dirty, the mist extractor will frequently plug and it may be removed. Subsequent scrubbers may need to be added or bypass lines installed to allow equipment to be cleaned with relative ease.

Horizontal separators are most often efficient for large volumes of total fluids and when large amounts of dissolved gases are present with the liquid. The greater liquid surface area provides for optimum conditions for releasing gas from the liquid.

Spherical separators are rarely used in production operations.

4.4 VERTICAL SEPARATORS

The following calculations are presented as a guide to the design and sizing of two-phase separators. Sizing should be based on the maximum expected instantaneous rate.

For practical purposes, for vertical separators, Equation 4.2 is written as

$$V_t = K \left(\frac{\gamma_L - \gamma_g}{\gamma_g} \right)^{0.5} \qquad (4.7)$$

where V_t = terminal velocity of liquid droplets or maximum allowable
superficial velocity of gas in ft./s
K = a constant depending on design and separating conditions
in ft./s. See Table 4.6 [11].

Example 4.1
A vertical gravity separator is required to handle 10 MMscfd of 0.6 specific gravity gas at an operating pressure of 1,000 psia and a temperature of 60°F. Liquid flow = 2,000 bpd of 40° API oil, μ_g = 0.014 cp.

Solution
1. Find K from Table 4.6 where K = 0.26 ft./s.

TABLE 4.6 Typical K-Factor Values in Equation 4.7

Separator Type	K factor (ft./s)
Horizontal (w/vertical pad)	0.40 to 0.50
Vertical or Horizontal (w/horizontal pad)	0.18 to 0.35
@Atm. pressure	0.35
@300 psig	0.33
@600 psig	0.30
@900 psig	0.27
@1500 psig	0.21
Spherical	0.20 to 0.35
Wet steam	0.25
Most vapors under vacuum	0.20
Salt and caustic evaporators	0.15

1. $K = 0.35$ @ 100 psig (subtract 0.01 for every 100 psi above 100 psig).
2. For glycol and amine solutions, multiply K by 0.6 to 0.8.
3. Typically use $\frac{1}{2}$ of the above K values for approximate sizing of vertical separators without woven wire de-misters.
4. For compressor suction scrubbers and expander inlet separators multiply K by 0.7 to 0.8.

2. Calculate minimum diameter, such that if

$$V_t = K \left(\frac{\gamma_L - \gamma_g}{\gamma_g} \right)^{0.5}$$

then

$$V_g = \frac{Q_A}{A_g} \frac{\text{ft.}^3/\text{s}}{\text{ft.}^2}$$

$$A_g = \frac{\pi}{4} D^2$$

Gas-specific weight is $\gamma_g = (PM/ZRT)$

$M = 0.6 \times 29 = 17.4$

$P = 1,000 \, \text{psia}$

$T = 520°R$

$R = 10.73 \, \text{ft.}^3 \cdot \text{psia} \cdot 1b_{mol}^{-1}, R^{-1} Z = 0.84$ (for given P, T, M)

$$\gamma_g = \frac{1,000 \times 17.4}{0.84 \times 10.73 \times 520} = 3.713 \, \text{lb/ft.}^3$$

Liquid-specific weight is

$$\gamma_L = 62.4 \times SG_0 = 62.4 \frac{141.5}{131.5 + 40} = 51.48 \, \text{lb/ft.}^3$$

Weight rate of flow w (lb/s) is

$$\dot{w} = \frac{Q(scf)17.4(lb_m/lb\,mol)}{379.4(scf/lb\,mol) \times (24 \times 3{,}600)s}$$

$$= \frac{10 \times 10^6 \times 17.4}{379.4 \times 86{,}400} = 5.3\,lb/s$$

$$Q_A = \frac{w}{\gamma_g} = \frac{5.3\,lb_m/s}{3{,}713\,lb_m/ft.^3} = 1.43\,ft.^3/s$$

$$V_g = \frac{Q_A}{A_g} = \frac{143 \times 4}{\pi D^2} = \frac{1.82}{D^2}$$

$$V_t = V_g$$

$$\frac{1.82}{D^2} = 0.26\left(\frac{51.48 - 3.713}{3.713}\right)^{0.5} = 0.99\,ft./s$$

$$D = 1.39\,ft. = 16.76\,in.$$

Minimum diameter is 20 in. (see Table 4.3). If

$$V_t = \left[\frac{4gD_p(\gamma_L - \gamma_g)}{3\gamma_g C_D}\right]^{0.5}$$

D_P is usually 100 to 150 μm.
Assume $D_P = 150\,\mu m = 150 \times 0.00003937/12 = 4.92 \times 10^{-6}$ ft.
From Equation 4.4

$$C_D(Re)^2 = \frac{0.95 \times 10^8 \times 3.713(0.000492)^3(51.48 - 3.713)}{(0.014)^2}$$

$$= 10{,}238$$

From Figure 4.5

$$C_D = 0.99$$

$$V_t = \left(\frac{4 \times 32.174 \times 0.000492(51.48 - 3.713)}{(0.014)^2}\right)^{0.5}$$

$$= 0.52\,ft./s$$

$$V_g = \frac{1.82}{D^2} = 0.52\,ft./s$$

$D = 1.87\,ft. = 22.5\,in.$, and from Table 4.3, the minimum diameter is 24 in.

Assume $D_p = 100\,\mu m = 100 \times 0.00003937/12 = 0.000328\,ft.$

$$C_D(Re)^2 = \frac{0.95 \times 10^8 \times 3.713(0.000328)^3(51.48 - 3.713)}{(0.014)^2}$$

$$= 2,592$$

From Equation 4.5

$$C_D = 1.75$$

$$V_t = \frac{(4 \times 32.174 \times 0.000328 \times 48.307)}{3 \times 3.713 \times 1.75} = 0.1046\,ft./s$$

$$V_g = \frac{1.82}{D^2} = 0.1046\,ft./s$$

$$D = 4.17\,ft. = 50\,in.$$

The maximum allowable superficial velocity calculated from the factors in Table 4.6 is for separators normally having a wire mesh mist extractor (Figure 4.7). This rate should allow all liquid droplets larger than 10 µm to settle out of the gas. The maximum allowable superficial velocity should be considered for other types of mist extractors.

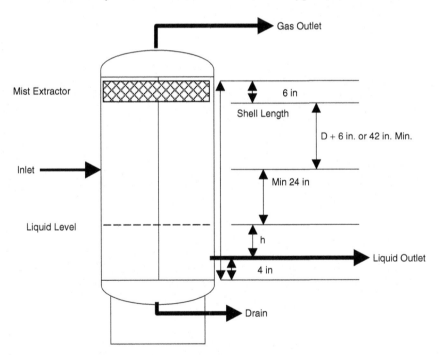

FIGURE 4.7 The geometry of a vertical separator [15].

Further calculations refer to $D = 20$ in. (separator with a mist extractor). Diameters $D = 24$ in. and $D = 50$ in. refer to separator without a mist extractor.

3. Calculate liquid level h. A certain liquid storage is required to ensure that the liquid and gas reach equilibrium at separator pressure. This is defined as "retention time" where liquid is retained in the vessel assuming plug flow. The retention time is thus the volume of the liquid storage in the vessel divided by the liquid flowrate.

Basic design criteria for liquid retention time in two-phase separators are generally as follows [17]:

Oil Gravities	Minutes (Typical)
Above 35° API	1
20–30° API	1 to 2
10–20° API	2 to 4

$$t = \frac{\text{Vol. ft.}^3}{Q \ \text{ft.}^3/s}$$

$$\text{Vol.} = \frac{\pi D^2 h}{4 \ 12} = \frac{\pi d^2 h}{4(144)12} = 4.545 \times 10^{-4} d^2 h$$

$$D = 20 \text{ in.}$$

$$\text{Vol.} = 0.1818h$$

$$Q_L = \text{bpd}$$

$$Q = Q_L \times 5.615 \left(\frac{\text{ft.}^3}{\text{barrel}} \right) \times \left(\frac{\text{day}}{24 \text{hr}} \right) \times \left(\frac{\text{hr}}{3,600 \text{s}} \right) = 0.000065 Q_L$$

$$t = \frac{0.1818h}{0.000065 Q_L} = 2,797 \frac{h}{Q_L}$$

Assume $t_r = 1 \text{ min}$ (API $= 40°$)
So,

$$h = \frac{t_r Q_L}{2,797} = \frac{2,000}{2,797} = 0.715 \text{ft.} = 8.6 \text{in.}$$

4. Calculate seam-to-seam length (L_{ss}). The seam-to-seam length of the vessel should be determined from the geometry of the vessel once a diameter and weight of liquid volume are known.

$$L_{ss} = \frac{h + 76}{12} \text{(in.)} \quad \text{or} \quad L_{ss} = \frac{h + D + 40}{12}$$

$$= \frac{8.6 + 76}{12} = 7.05 \text{ft.}$$

5. Compute slenderness ratio (L_{ss}/D), which is usually in the range 3 to 4, such that

$$L_{ss}/D = \frac{7.05 \times 12}{20} = 4.23$$

6. Choosing the diameter because (L_{ss}/D) > 4, let's assume higher diameter $D = 24$ (next higher diameter from Table 4.3).

$$Vol = 0.2618h$$

$$t = \frac{0.2618h}{0.000065Q_L} = 4{,}028\frac{h}{Q_L}$$

If

$$t = 1\,min$$

then

$$h = \frac{2{,}000}{4{,}028} = 0.5\,ft. = 6\,in.$$

$$L_{ss} = \frac{6 + 76}{12} = 6.83\,ft. = 82\,in.$$

$$\frac{L_{ss}}{D} = \frac{82}{24} = 3.4$$

Proper size of separator is 24 in. $\times 7\frac{1}{2}$ ft.

4.5 HORIZONTAL SEPARATOR

In the case of horizontal separators, the gas drag force does not directly oppose the gravitational settling force. The thru droplet velocity is assumed to be the vector sum of the vertical terminal velocity and the horizontal gas velocity. The minimum length of the vessel is calculated by assuming the time for the gas to flow from the inlet to the outlet is the same as for the droplets to fall from the top of the vessel to the surface of the liquid. In calculating the gas capacity of horizontal separators, the cross-sectional area of that portion of the vessel occupied by liquid (at maximum level) is subtracted from the total vessel cross-sectional area, as shown in Figure 4.8.

Separators can be any length, but the ratio L_{ss} to D of a vessel is usually in the range of 2:1 to 4:1, so that

$$V_t = K\left(\frac{\gamma_L - \gamma_g}{\gamma_g}\right)\left(\frac{L}{10}\right)^{0.56} \quad (ft./s) \tag{4.8}$$

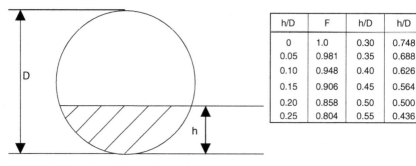

h/D	F	h/D	h/D
0	1.0	0.30	0.748
0.05	0.981	0.35	0.688
0.10	0.948	0.40	0.626
0.15	0.906	0.45	0.564
0.20	0.858	0.50	0.500
0.25	0.804	0.55	0.436

FIGURE 4.8   The fraction of the total area available for gas flow in a horizontal separator.

Sometimes separators without mist extractors are sized using Equation 4.7 with a constant K of typically 1/2 of that used for vessels with mist extractors.

Example 4.2

Solve Example 4.1 in the case of horizontal separator with mist extractor if $F = 0.564$ and 0.436

1. Find K value from Table 4.6 where $K = 0.3$
2. The gas capacity constraint is determined from

$$V_g = \frac{Q}{A_g}$$

$$A_g = 0.564 \times \frac{\pi}{4} D^2 = 0.443 D^2$$

$$Q = Q_g \times \frac{10^6 \text{scf}}{\text{MMscf}} \times \frac{\text{day}}{24\,\text{hr}} \times \frac{\text{hr}}{3{,}600\,\text{s}} \times \frac{14.7}{P} \times \frac{TZ}{520} = 0.327 \frac{TZ}{P} Q_g$$

$$V_g = \frac{0.327 TZ Q_g}{0.443 D^2 P} = 0.74 \frac{TZ Q_g}{D^2 P}$$

The residence time of the gas (t_g) has to be equal to the time required for the droplet to fall to the gas–liquid interface t_r:

$$t_g = \frac{L_{eff}}{V_g} \frac{\text{Effective length for separator}}{\text{Gas velocity}} = \frac{L_{eff} D^2 P}{0.74 TZ Q_g}$$

$$t_d = \frac{D - h}{V_t}$$

If $F = 0.564$, then $h/D = 0.45$, where $h = 0.45D$ and $t_d = 0.55D/V_t$. If $L_{ss} = L_{eff} + D$ for gas capacity, then

$$V_t = K \left(\frac{\gamma_L - \gamma_g}{\gamma_g} \right)^{0.5} \left(\frac{L_{eff} + D}{10} \right)^{0.56}$$

and

$$t_d = \frac{0.55D}{0.3 \left(\dfrac{\gamma_L - \gamma_g}{\gamma_g} \right)^{0.5} \left(\dfrac{L_{eff} + D}{10} \right)^{0.56}}$$

where $t_g = t_d$ and $\dfrac{L_{eff}D^2P}{0.74TZQ_g} = \dfrac{0.55D}{0.3 \left(\dfrac{\gamma_L - \gamma_g}{\gamma_g} \right)^{0.5} \left(\dfrac{L_{eff} + D}{10} \right)^{0.56}}$

$$D^2 L_{eff} = \frac{6D}{(L_{eff} + D)^{0.56}}$$

From the above constraint for given D, L_{eff} could be calculated, but gas capacity does not govern.

3. Liquid capacity constraint is calculated from

$$t_r = \frac{\text{Vol.}}{Q}$$

where $\text{Vol.} = (\pi D^2 / 4)(1 - F) L_{eff} = 0.343 D^2 L_{eff}$

$$Q = Q_L \times 5.615 \,(\text{ft.}^3/\text{barrel}) \times (\text{day}/24\,\text{hr})$$
$$\times (\text{hr}/3,600\,\text{s}) = 0.000065\,Q_L$$

$$t_r = 0.343 D^2 L_{eff}/0.000065\,Q_L$$

$$D^2 L_{eff} = (t_r \times 0.000065\,Q_L)/0.343$$

If t_r is in minutes, then $D^2 L_{eff} = 0.01137\,t_r Q_L$.

4. Assume retention time $t_r = 1\,\text{min}$. Compute combination D (Table 4.4) and L_{eff}, such that

$$D = 2\,\text{ft.} \quad L_{eff} = 5.7\,\text{ft.} \quad L_{ss} = \tfrac{4}{3} L_{eff} = 7.6\,\text{ft.}$$

$$D = 2.5\,\text{ft.} \quad L_{eff} = 3.6\,\text{ft.} \quad L_{ss} = \tfrac{4}{3} L_{eff} = 4.9\,\text{ft.}$$

5. Compute slenderness $(L_{ss}/D) = 3.8$ and 1.96, respectively. $D = 2\,\text{ft.}$ (24 in.) is a proper size.

A second case is where $F = 0.436$. The gas capacity constant is determined from

$$V_g = \frac{Q}{A_g} \quad \text{where} \quad Q = 0.327\frac{TZQ_g}{P}$$

$$A_g = 0.436 \times \frac{\pi}{4}D^2 = 0.342D^2$$

$$V_g = \frac{0.327TZQ_g}{0.342D^2P} = 0.96\frac{TZQ_g}{D^2P}$$

$$t_g = \frac{L_{eff}}{V_g} = \frac{L_{eff}D^2P}{0.96TZQ_g}$$

$$t_d = (D-h)/V_t \quad \text{if} \quad F = 0.436, \quad \frac{h}{D} = 0.55, \quad h = 0.55D$$

$$t_d = \frac{0.45D}{V_t} \quad \text{if} \quad V_t = k \times 3.587\left(\frac{L_{ss}}{10}\right)^{0.56}$$

$L_{ss} = L_{eff} + D$ for gas capacity, such that

$$V_t = 0.988K\,(L_{eff} + D)^{0.56} \quad \text{where} \quad K = 0.3$$

$$t_d = \frac{1.52D}{(L_{eff} + D)^{0.56}}$$

$$t_g = t_d$$

$$\frac{L_{eff}D^2P}{0.96TZQ_g} = \frac{1.52D}{(L_{eff} + D)^{0.56}}$$

Substituting Q_g, P, T, Z gives

$$L_{eff}D^2 = \frac{6.4D}{(L_{eff} + D)^{0.56}}$$

Liquid capacity constraint is determined from

$$t_r = \frac{Vol.}{Q}$$

where Vol. $= (\pi D^2/4)\,(1-F)\,L_{eff} = 0.443D^2L_{eff}$

$Q = 0.000065\,Q_L$ if t_r is in min

$t_r = 0.443D^2L_{eff}/0.000065\,Q_L$

$D^2L_{eff} = 0.0088\,t_r\,Q_L$

Assume retention time $t_r = 1$ min and $D = 2$ ft.

$$L_{eff} = \frac{0.0088 \times 1 \times 2,000}{4} 4.4 \, ft.$$

$$L_{ss} = \frac{4}{3} \times 4.4 = 5.9 \, ft.$$

$$D = 2.5$$

$$L_{eff} = \frac{0.0088 \times 1 \times 2,000}{2.5 \times 2.5} = 2.8 \, ft.$$

$$L_{ss} = \frac{4}{3} \times 2.8 = 3.8 \, ft.$$

As you can see, the liquid level has significant meaning for separator length. Usually, the fraction of the total area F available for gas flow is equal to 0.5. The liquid level control placement of a horizontal separator is more critical than in a vertical separator and the surge space is somewhat limited.

4.6 VESSEL INTERNALS [18]

The proper selection of internals can significantly enhance the operation of separators. Proprietary internals often are helpful in reducing liquid carryover at design conditions. Nevertheless, they cannot overcome an improper design or operation at off-design conditions.

Production equipment involving the separation of oil and gas often uses impingement-type mist-extraction elements. This element is usually of the vane type or of knitted wire.

The vane type consists of a labyrinth formed with parallel metal sheets with suitable liquid collection "pockets." The gas, in passing between plates, is agitated and has to change direction a number of times. Obviously, some degree of centrifugation is introduced, for as the gas changes direction the heavier partials tend to be thrown to the outside and are caught in the pockets provided.

Coalescence of small particles into those large enough to settle by gravity is provided by two mechanisms: agitation and surface. The surface of the element is usually wet, and small particles striking it are absorbed. Inasmuch as the pockets are perpendicular to the gas flow, the liquid thus formed does not have to flow against the gas. Consequently, small compact units have a large capacity.

As the plates are placed closer together and more pockets are provided, greater agitation, centrifugal force, and collection surface are provided, but the pressure drop is increased correspondingly. Thus, for a given flow rate, the collection efficiency is normally some function of the pressure drop.

In the average application, this pressure drop varies from 1.2 to 10 in. (3 to 25 cm) of water. Because of this pressure drop and to prevent gas bypassing the extractor, a liquid-collection pas incorporating a liquid seal is necessary for the liquid to drain properly.

Increased use has been made of mist extractors composed of a knitted wire mesh supported on a lightweight support. This material has given generally favorable results and has a low installed cost.

The element consists of wire knitted into a pad having a number of unaligned, asymmetrical openings. Although similar in appearance to filter media, its action is somewhat different. The latter are rather dense and have small openings. This knitted wire, on the other hand, has about 97% to 98% free voids and collects the particles primarily by impingement.

The material is available in single wound units of varying thicknesses in diameter up to 35 in. (90 cm) or in laminated strips for insertion through manholes in large process vessels.

The principle of separation is similar to that of the vane-type unit. The gas flowing through the pad is forced to change direction a number of times, although centrifugal action is not so pronounced. Impingement is the primary mechanism.

A liquid particle striking the metal surface, which it does not "wet," flows downward where adjacent wires provide some capillary space. At these points, liquid collects and continues to flow downward. Surface tension tends to hold these drops on the lower face of the pad until they are large enough for the downward force of gravity to exceed that of the upward gas velocity and surface tension.

Efficiency is a function of the number of targets presented. This may be accomplished by increasing the pad thickness, changing wire diameter, or the closeness of the weave.

The wire mesh normally used falls within the following range:

- Wire diameter—0.003 to 0.011 in. (0.0076 to 0.028 cm)
- Void volume—92% to 99.4%
- Specific weight—3 to 33 lb/ft.^{3} (48 to 529 kg/m^3)
- Surface area—50 to 600 ft.^{2}/ft.^{3} (164 to 1970 m^2/m^3)

The most commonly used wire has a void volume of 97% to 98%, a bulk-specific weight approximately 12 lb/ft.^{3} (192 kg/m^3), a surface area of 100 to 125 ft.^{2}/ft.^{3} (328 to 410 m^2/m^3), with a wire diameter of 0.011 in. (0.028 cm). A pad thickness of 4 to 6 in. (10 to 15 cm) is sufficient for most separator applications, although thicknesses up to 3 ft. (0.9 m) have been reported. In separator service, 4 to 6 in. (10 to 15 cm) will normally suffice.

Any common metal may be used in these units, including carbon steel, SS, aluminum, monel, etc. The pressure drop is a function of the entrainment load, the pad design, and gas velocity but will not exceed 3 cm of water in the average insulation. Because of this small pressure drop, the elements do not have to be "held down" and are normally only wired to the support grid to prevent shifting unless surging flow is anticipated.

Experience has shown that the support grid should contain at least 90% free area in order to eliminate any restrictions to liquid drainage. The pads are light in weight so that a light angle-iron support is adequate. When both liquid and solids are present, a portion of the latter obviously will be scrubbed out. If only dry solids are present, the efficiency of this design is substantially less. Many vessel carryover problems are encountered. Foaming is a major culprit and requires more than simply better mist extraction. Most such problems develop by default. The vendor automatically uses the standard sizing curves and equipment, and the buyer assumes this will be good enough.

With glycols, amines, and similar materials, which tend to foam, a dual mist extractor would normally be specified—the lower one being of the vane type and the upper one being a wire mesh. A space of 6 to 12 in. (15 to 30 cm) would be left between them. The vane type will handle large volumes of liquid but is relatively inefficient on small droplets. It, therefore, serves as a bulk removal device (and helps to coalesce foam). The wire mesh, which has limited liquid capacity, may therefore operate more effectively.

When using the vane-type mist extractor, one must be careful that the pressure drop across it does not exceed the height above the liquid level if a downcomer pipe is used. Otherwise, liquid will be "sucked out" overhead. The downcomer pipe can become partially plugged to accentuate the problem. Two wire mesh pads may be used in like fashion, with the first being used as a coalescer. As a rule of thumb, the coalescer pad should have about half the free space area of the second pad. Any wire mesh pad should be installed so that the flow is perpendicular to the pad face (pad is horizontal in a vertical vessel).

With materials like glycol and amine, which wet metal very well, Teflon-coated mesh may prove desirable. Remember, the liquid must be nonwetting in order to stay as droplets that can run down the wires and coalesce into bigger droplets. A wetting fluid will tend to "run up" the wires. It has been noted that centrifugal force is an integral part of the separation process. The standard oil and gas separator may have an inlet that utilizes centrifugal force to separate the larger droplets. The same principle is used in some mist extractor elements except that higher velocities are needed to separate the smaller droplets. The velocity needed for separation is a function of the particle diameter, particle and gas specific weight and the gas viscosity.

With a given system, the size of particle collected is inversely proportional to the square root of the velocity. Consequently, the success of a cyclonic mist extractor is dependent upon the velocity attained. Furthermore, the velocity needed to separate a given size of particle must increase, as the density of the particle becomes less. In addition to producing the necessary velocity, the mist extractor must provide an efficient means of collecting and removing the particles collected in order to prevent re-entrainment.

One common type of equipment is often called a *steam separator* because it has been widely used to separate condensate and pipe scale in steam systems. Normally, a relatively small vessel imparts a high velocity to the incoming gas and then makes the gas change direction radically to prevent re-entrainment. In general, it will separate particles 40 microns and larger very efficiently.

Another type uses the same principle but, in addition, forces the gas to pass through a labyrinth that introduces impingement effects and forces the gas to change direction a number of times. This is, in reality, a combination type and is relatively efficient. The general performance characteristics are the same as efficient mist extractors of other types. Some, however, are complex and relatively expensive.

4.7 OIL–WATER–GAS SEPARATION

Two liquids and gas or oil–water–gas separation can be easily accomplished in any type of separator by installing either special internal baffling to construct a water lap or crater siphon arrangement or by use of an interface liquid level control. A three-phase feature is difficult to install in a spherical separator due to the limited internal space available. With three-phase separation, two liquid level controls and two liquid dump valves are required. Figures 4.9 to 4.11 [15] illustrate schematics of some three-phase

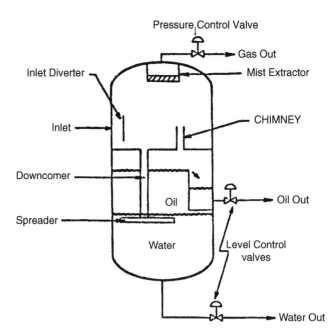

FIGURE 4.9   Three-phase vertical separator.

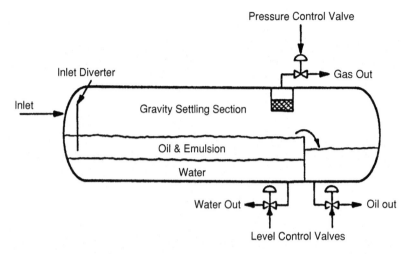

FIGURE 4.10 Three-phase horizontal separator.

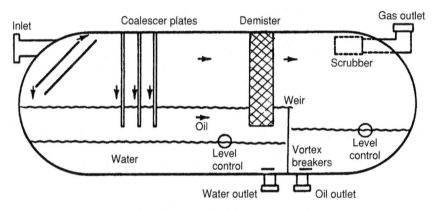

FIGURE 4.11 Three-phase horizontal separator with coalescer plates.

separators. The basic design aspects of three-phase separation have been covered under "Sizing of Two Phase Oil–Gas Separators". Regardless of shape, all three-phase vessels must meet the following requirements:

- Liquid must be separated from the gas in a primary separating section.
- Gas velocity must be lowered to allow liquids to drop out.
- Gas is then scrubbed through an efficient mist extractor.
- Water and oil must be diverted to a turbulence-free section of the vessel.
- Liquids must be retained in the vessel long enough to allow separation.
- The water–oil interface must be maintained.
- Water and oil must be removed from the vessel at their respective outlets.

TABLE 4.7   Typical Retention Time for Liquid-Liquid Separation.

Type of Separation	Retention Time
Hydrocarbon/water separators	
Above 35° API hydrocarbons	3 to 5 min
Below 35° API hydrocarbons	
100°F and above	5 to 10 min
80°F	10 to 20 min
60°F	20 to 30 min
Ethylene glycol/hydrocarbon separators (cold separators)	20 to 60 min
Amine/hydrocarbon separators	20 to 30 min
Coalescers, hydrocarbon/water separators	
100°F and above	5 to 10 min
80°F	10 to 20 min
60°F	20 to 30 min
Caustic/propane	30 to 45 min
Caustic/heavy gasoline	30 to 90 min

Sizing a three-phase separator for water removal is mainly a function of retention time; required retention time is related to the volume of the vessel, the amount of liquid to be handled and the relative specific gravities of the water and oil. The effective retention volume in a vessel is that portion of the vessel in which the oil and water remain in contact with one another. As far as oil–water separation is concerned, once either substance leaves the primary liquid section, although it may remain in the vessel in a separate compartment it cannot be considered as part of the retention volume. There are two primary considerations in specifying retention time:

- Oil settling time to allow adequate water removal from oil
- Water settling time to allow adequate oil removal from water

Basic design criteria for liquid retention time in three-phase separators are given in Table 4.7 [11].

Example 4.3
Determine the size of a vertical separator to separate 6,300 bpd of oil from its associated gas, water 500 bpd (SG = 0.75) at a pressure of 300 psig, and a temperature with $SG_g = 1.03$, 100°F. The oil has a density of 49.7 lb/ft.^{3} and solution gas/oil ratio of 580 scf/STB at 60°F and 14.7 psia.

1. The gas capacity constraint $V_g = V_t$, such that

$$V_g = \frac{Q}{A_g} \quad Q = \frac{R_s \times Q_0}{86,400} \times \frac{14.7}{P} \times \frac{TZ}{520}$$

$P = 300 \quad T = 560 \quad Z = 0.94$ (from chart)

$Q = 2.1\,\text{ft.}^3/\text{s}$

$$A_g = \frac{\pi D^2}{4} = 0.785 D^2$$

$$V_t = K \left(\frac{\gamma_0 - \gamma_g}{\gamma_g} \right) \quad K = 0.33 \text{ (from Table 4.6)}$$

$\gamma_0 = 49.7\,\text{lb/ft.}^3$

$$\gamma_g = \frac{PM}{ZRT} = \frac{300 \times 29 \times 0.75}{0.94 \times 10.73 \times 560} = 1.155\,\text{lb/ft.}^3$$

$$\left(\frac{\gamma_0 - \gamma_g}{\gamma_g} \right)^{0.5} = \left(\frac{49.7 - 1.155}{1.155} \right)^{0.5} = 6.48$$

$V_t = 2.14\,\text{ft./s}$

$D = 1.117\,\text{ft.} = 13.4\,\text{in.}$

2. Calculate minimum diameter from requirement for water droplets to fall through oil layer. Use 500-μm droplets if no other information is available.
   For $\mu_0 =$ assume 3.0 cp.

$$V_t = V_0$$

According to Stokes law

$$V_t = 2660 \frac{(\gamma_w - \gamma_0)}{\mu} D_p^2$$

$V_t(\text{ft./s}); \quad \gamma_w, \gamma_0 (\text{lb/ft.}^3); \quad D_p(\text{ft.}); \quad \mu(\text{cp})$

$$V_0 = \frac{Q}{A} \quad Q(\text{ft./s}); \quad A(\text{ft.}^2); \quad Q_0(\text{bpd})$$

$Q = 6.49 \times 10^{-5} Q_0$

$$A = \frac{\pi D^2}{4} = 0.785 D^2 \quad \text{Where D is in ft.}$$

$$V_0 = \frac{6.49 \times 10^{-5} Q_0}{0.785 D^2} = 8.27 \times 10^{-5} \frac{Q_0}{D^2}$$

$$8.27 \times 10^{-5} \frac{Q_0}{D^2} = 2660 \left(\frac{\gamma_w - \gamma_0}{\mu_0} \right) D_p^2$$

$D_p = 500\,\mu\text{m} = 500 \times 3.2808 \times 10^{-6}\,\text{ft.}$

$$D_p = 1.64 \times 10^{-3}\,\text{ft}.$$

$$D = \left[\frac{8.27 \times 10^{-5} Q_0 \mu_0}{2600\,(\gamma_w - \gamma_0)\, D_P^2} \right]^{0.5} = 4.15\,\text{ft}. = 49.8\,\text{in}.$$

3. Calculate the total weight of oil and water in separator $(h_0 + h_w)$

$$t_r = \text{Vol.}/Q$$

$$\text{Vol.} = (\pi D^2/4) h$$

$$Q = 6.49 \times 10^{-5}\, Q_L \quad \text{If t is in min}$$

$$t_r\, 60 = \frac{\pi D^2 h}{4 \times 6.49 \times 10^{-5} Q_L} \qquad Q_L\,(\text{bpd})$$

$$D^2 h = 4.958 \times 10^{-3} Q_L t_r$$

For two-phase separator design

$$D^2 h_0 = 4.958 \times 10^{-3} Q_0 (t_r)_0$$

$$D^2 h_w = 4.958 \times 10^{-3} Q_w (t_r)_w$$

$$h_0 + h_w = 4.958 \times 10^{-3} \frac{Q_0 (t_r)_0 + Q_w (t_r)_w}{D^2}$$

where $h_0 =$ height of oil pad in ft.
$h_w =$ height from water outlet to interface in ft.
Choose a nominal diameter from Table 4.3, considering minimum diameter in pts 2.54 in (4.5 ft.) is a proper choice. $(t_r)_0 = (t_t)_w = 5$ min from Table 4.7.

$$h_0 + h_w = 4.958 \times 10^{-3} \frac{500 \times 5 + 6{,}300 \times 5}{(4.5)^2} = 8.32\,\text{ft}.$$

$$\text{If } D = 5\,\text{ft}. \quad h_0 + h_w = 6.74\,\text{ft}.$$

4. Calculate seam-to-seam length and slenderness ratios for $D = 4.5$ ft.

$$L_{ss} = h_0 + h_w + \frac{76}{12} = 8.32 + 6.33 = 14.6\,\text{ft}.$$

$$\frac{L_{ss}}{D} = \frac{14.6\,\text{ft}}{4.5} = 3.2 \quad \text{for} \quad D = 5\,\text{ft}.$$

$$L_{ss} = 6.74 + \frac{76}{12} = 13\,\text{ft}.$$

$$\frac{L_{ss}}{D} = \frac{13\,\text{ft.}}{5} = 2.6$$

Vertical three-phase separators have slenderness ratios on the order of 15 to 3. In this case, both diameters are acceptable.

Example 4.4

Determine the size of a horizontal separator for data given in Example 4.3 and $F = 0.5$.

1. Gas capacity constraint: $t_g = t_d$

$$V_g = \frac{Q}{A_g}$$

$$Q = \frac{R \times Q_0}{86,400} \times \frac{14.7}{P} \times \frac{TZ}{520} = 2.1 \, \text{ft.}^3/\text{s}$$

$$A_g = \frac{\pi D^2}{4} \times F = 0.393 D^2$$

$$V_g = \frac{2.1 \, \text{ft.}^3/\text{s}}{0.393 D^2} = \frac{5.34}{D^2}$$

$$t_g = \frac{L_{eff}}{V_g} = \frac{L_{eff} D^2}{5.34}$$

$$t_d = \frac{D/2}{V_t} = \frac{D}{2V_t}$$

$$V_t = K \left(\frac{\gamma_L - \gamma_g}{\gamma_g} \right)^{0.5} \left(\frac{L_{ss}}{10} \right)^{0.56}$$

$$= 0.33 \left(\frac{49.7 - 1.155}{1.155} \right)^{0.5} \left(\frac{L_{eff} + D}{10} \right)^{0.56}$$

$$= 0.589 \, (L_{eff} + D)^{0.56}$$

$$\frac{L_{eff} D^2}{5.34} = \frac{D}{0.589 \, (L_{eff} + D)^{0.56}}$$

$$L_{eff} D^2 = \frac{9.07 D}{(L_{eff} + D)^{0.56}}$$

For assumed D, calculate L_{eff}:

D(ft.)	L_{eff}(ft.)
4.0	0.93
4.5	0.79
5.0	0.72

2. Calculate maximum oil pad thickness $(h_0)_{max}$. Use 500 μm droplet if no other information is available $t_w = t_0$:

$$t_w = \frac{h_0}{V_t} \quad V_t = 2{,}660\frac{(\gamma_w - \gamma_0)}{\mu}D_p^2$$

$$t_w = \frac{h_0\mu}{2{,}660\,(\gamma_w - \gamma_0)\,D_p^2}$$

$$t_0 = 60(t_r)_0 \quad t_r(min)$$

$$\frac{h_0\mu}{2{,}660\,(\gamma_w - \gamma_0)\,D_p^2} = 60(t_r)_0$$

$$h_0 = \frac{159{,}600\,(\gamma_w - \gamma_0)\,D_p^2\,(t_r)_0}{\mu}$$

This is the maximum thickness the oil pad can be and still allow the water droplets to settle out in time $(t_r)_0$. If $D_p = 500\,\mu m = 500 \times 3.2808 \times 10^{-6}$ ft. $= 1.64 \times 10^{-3}$ ft.

$$h_0 = \frac{0.43\,(\gamma_w - \gamma_0)\,(t_r)_0}{\mu}$$

For a given retention time (from Table 4.7, is equal to 5 min) and a given water retention time (also 5 min), the maximum oil pad thickness constraint establishes a maximum diameter.

$$\gamma_w = 64.3\,lb/ft.^3 \quad \gamma_0 = 49.7\,lb/ft.^3 \quad \mu = 3\,cp$$

$$(h_0)_{max} = 10.46\,ft.$$

3. Calculate the fraction of the vessel cross-sectional area occupied by the water phase (see Figures 4.8 and 4.12) A, A_w, A_0 (ft.²):

$$Q(ft.^3/s) \quad t(s) \quad L_{eff}(ft.)$$

$$A = \frac{Qt}{L_{eff}}$$

$$Q = 6.49 \times 10^{-5}Q_0, \quad also \; Q = 6.49 \times 10^{-5}Q_w$$

$$t_0 = 60(t_r)_0 \quad t = 60(t_r)_w$$

$$A_0 = 3.89 \times 10^{-3}\frac{Q_0\,(t_r)_0}{L_{eff}} \quad A_w = 3.89 \times 10^{-3}\frac{Q_w\,(t_r)_w}{L_{eff}}$$

$$A = 2(A_0 + A_w)$$

Hence

$$\frac{A_w}{A} = \frac{3.89 \times 10^{-3}Q_w\,(t_r)_w\,L_{eff}}{2L_{eff}3.89 \times 10^{-3}\,[Q_w\,(t_r)_w + Q_0\,(t_r)_0]}$$

$$= \frac{0.5Q_w\,(t_r)_w}{Q_w\,(t_r)_w + Q_0\,(t_r)}$$

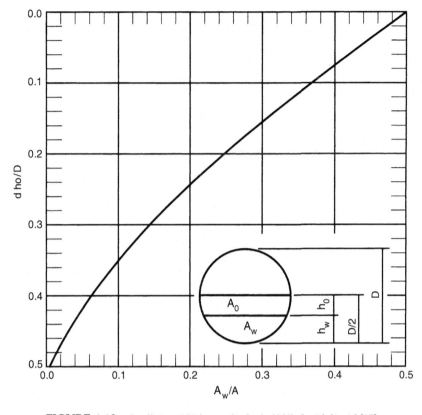

FIGURE 4.12 Coefficient "d" for a cylinder half filled with liquid [15].

4. From Figure 4.12, determine the coefficient "d," and calculate D_{max} for oil pad thickness constraint:

$$\frac{A_w}{A} = \frac{0.5(5)500}{500(5) + 6,300(5)} = 0.037$$

$$d = \frac{h_0}{D} = 0.43$$

$$D_{max} = \frac{(h_0)_{max}}{d} = \frac{10.46}{0.43} = 24.3\,\text{ft.}$$

D_{max} depends on Q_0, Q_w, $(t_r)_0$, and $(t_r)_w$.

5. Liquid retention constraint:

$$t = \frac{\text{Vol.}}{Q}$$

$$\text{Vol.} = \frac{1}{2}\left(\frac{\pi D^2}{4}L_{eff}\right) = 0.393 D^2 L_{eff}$$

$$(\text{Vol.})_0 = 0.393 D^2 L_{eff}\left(\frac{A_0}{A_{liq}}\right)$$

$$(\text{Vol.})_w = 0.393 D^2 L_{eff}\left(\frac{A_w}{A_{liq}}\right)$$

Q_0 and Q_w are in bpd:

$$Q = 6.49 \times 10^{-5} Q_0 \quad Q = 6.49 \times 10^{-5} Q_w$$

$$t_0 = \frac{0.393 D^2 L_{eff}\left[\dfrac{A_0}{A_{liq}}\right]}{6.49 \times 10^{-5} Q_0}$$

$$\frac{Q_0 t_0}{D^2 L_{eff}} = 6,055 \frac{A_0}{A_{liq}}, \quad \text{also} \quad \frac{Q_w t_w}{D^2 L_{eff}} = 6,055 \frac{A_w}{A_{liq}}$$

$(t_r)_0$ and $(t_r)_w$ are in min:

$$\frac{Q_0 (t_0)_0}{D^2 L_{eff}} = 101 \frac{A_0}{A_{liq}} \quad \frac{Q_w (t_r)_w}{D^2 L_{eff}} = 101 \frac{A_w}{A_{liq}}$$

Adding by sides

$$\frac{Q_0 (t_r)_0 + Q_w (t_r)_w}{D^2 L_{eff}} = 101 \frac{A_0 + A_w}{A_{liq}} = 101$$

$$D^2 L_{eff} = 0.0099[Q_0(t_r)_0 + Q_w(t_r)_w]$$

for given $(t_r)_0$ and $(t_r)_w = 5\,\text{min}$, and

$$Q_0 = 6,300\,\text{bpd} \quad Q_w = 500\,\text{bpd}$$

$$D^2 L_{eff} = 0.0099[500(5) + 6,300(5)] = 336.6$$

6. Compute combinations of D and L_{eff}

$$L_{ss}, \quad \text{where} \quad L_{ss} = L_{eff}\frac{4}{3} \quad \text{and} \quad \frac{L_{ss}}{D}$$

D (ft.)	L_{eff}(ft.)	L_{ss}	L_{ss}/D
5.0	13.5	18.0	3.5
5.5	11.1	14.8	2.7
4.5	16.6	22.2	4.9

For three-phase horizontal separators, slenderness ratio is in the range of 3 to 5 usually, so D = 4.5 and 5 ft. are proper choices. As one can see, liquid retention constraint limits three-phase separator size; gas capacity, and oil pad thickness do not govern.

4.8 TWO-STAGE SEPARATION SYSTEMS

In high-pressure gas-condensate separation systems, it is generally accepted that a stepwise reduction of the pressure on the liquid condensate will appreciably increase the recovery of stock tank liquids. The calculation of the actual performance of the various separators in a multistage system can be made, using the initial wellstream composition and the operating temperatures and pressures of the various stages. Theoretically, three or four stages of separation would increase the liquid recovery over two stages: the net increase over two-stage separation will rarely pay out the cost of the second and/or third separator. Therefore, it has been generally accepted that two stages of separation plus the stock tank are the most optimum considered. The actual increase in liquid recovery for two-stage separation over single stage may vary from 3% to 15% (sometimes even more) depending upon the wellstream composition, operating pressures and temperatures.

The optimum high stage or first separator operating pressure is generally governed by the gas transmission line pressure and operating characteristics of the well. This will generally range in pressure from 600 to 1200 psia (41.4 to 82.7 bar). For each high or first-stage pressure, there is an optimum low-stage separation pressure that will afford the maximum liquid recovery. This operating pressure can be determined from an equation based on equal pressure ratios between the stages.

$$R = \left(\frac{P_1}{P_s}\right)^{0.5} \quad \text{or} \quad P_2 = \frac{P_1}{R} = P_s R^{n-1} \tag{4.9}$$

where R = pressure ratio
  n = number of stages
  P_1 = first-stage separator pressure in psia
  P_2 = second-stage separator pressure in psia
  P_s = stock tank pressure in psia

Figure 4.13 illustrates a schematic flow diagram of a typical high-pressure well production equipment installation [19]. The basic equipment is illustrated for two-stage separation of the high-pressure stream. From the wellhead, the high-pressure wellstream flows through a high-pressure separator and indirect heater gas production unit. In this unit the inlet stream is heated prior to choking to reduce the wellstream pressure to sales line pressure. This is done to prevent the formation of hydrates in the choke or

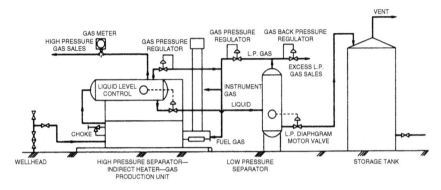

FIGURE 4.13   Two-stage separation system.

downstream of the choke in the separator or sales line. From the indirect heater, the wellstream passes to the high-pressure separator where the initial separation of the high-pressure gas stream and produced well fluids occur.

From the high-pressure separator, the gas flows through an orifice meter and to the sales gas line. The liquid from the high-pressure separator passes through a diaphragm valve where the pressure is reduced and it is discharged to a low-pressure flash separator. In the low-pressure flash separator that would operate at approximately 100 psi (6.9 bar), a second separation occurs between the liquids and the lighter hydrocarbons in the liquids. The gas released from the low-pressure flash separator is returned back to the high-pressure unit, where it may be used for both instrument and fuel gas for the indirect heater. As illustrated in Figure 4.13, a secondary makeup line is shown from the high-pressure separator, which would provide additional makeup gas for the instrument gas and fuel gas, if not enough gas was released from the low-pressure separator. However, typically more gas is released than is required, and the additional low-pressure gas may be sold in a low-pressure gas gathering system and/or used for other utility purposes, such as fuel for company compressor engines or other fired equipment on that lease. This may be for reboilers on dehydrators or acid gas sweetening units, etc. From the low-pressure flash separator, the liquid is discharged though another diaphragm valve into a storage tank that is generally operated at atmospheric pressure.

Example 4.5

Perform flash vaporization calculation to determine the increase recoveries that would be seen in both low-pressure flash gas as well as increased liquid recoveries in the storage tank. High-pressure gas $SG_g = 0.67$, the high-pressure separator pressures from 500 to 1,000 psi, the low-pressure separator $P = 100$ psi, and the storage tank pressure $P_s = 14.7$ psia. Temperature for all vessels is the same, 70°F.

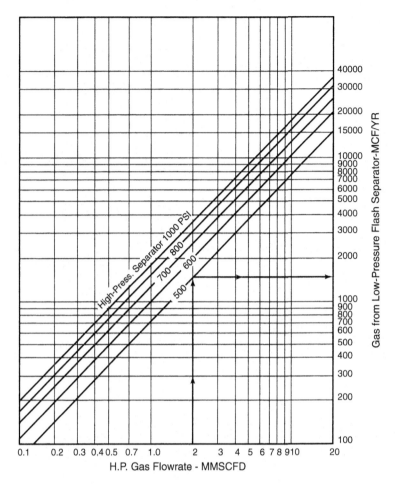

FIGURE 4.14 Low-pressure gas from flash separator [20].

Results

Figure 4.14 illustrates the gas produced from the low-pressure flash separator for the above-described wellstream at various high-pressure operating pressures (line pressure). The gas produced from the low-pressure flash separator in Mcf per year may be read from ordinate, based on the high-pressure gas stream flow rate in MMscfd and the high-pressure separator operating pressure.

Figure 4.15 illustrates the increase in stock tank liquid recovery that would be achieved using the low-pressure flash separator. This chart is also based on high-pressure gas flow-rate in MMscfd and the high-pressure separator operating pressure. The increase in stock tank liquid recovery may be read from the chart in bbl/year.

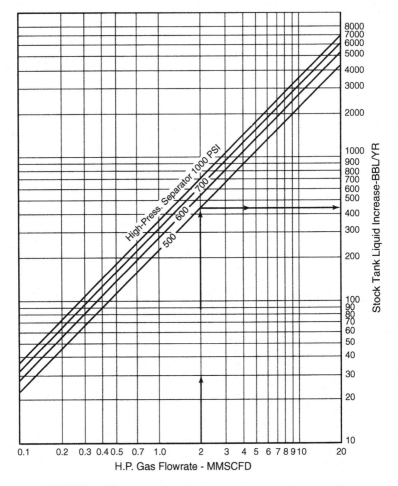

FIGURE 4.15   Stock tank liquid increase with flash separator [20].

This additional recovery not only gives profit but also prevents the unneeded waste of precious hydrocarbon energy that would normally be vented out the stock tank using only single-stage separation.

4.9 CRUDE OIL TREATING SYSTEMS

Water content of the untreated oil may vary from 1% to over 95%. Purchasers, depending on local conditions, accept a range from 0.2% to 3% of water in oil. When water forms a stable emulsion with the crude oil and cannot be removed in conventional storage tanks, emulsion-treating methods must be used.

An *emulsion* is a heterogeneous liquid system consisting of two immiscible liquids with one of the liquids intimately disposed in the form of droplets in the second liquid (the water remaining is less than 10% of the oil). A common method for separating water–oil emulsion is to heat the stream. The use of heat in treating crude oil emulsions has four basic benefits:

1. Heat reduces the viscosity of the oil, resulting in a greater force during collision of the water droplets.
2. Heat increases the droplets' molecular movement.
3. Heat can enhance the action of treating chemicals, causing the chemicals to work faster to break the film surrounding the droplets of the dispersed phase of the emulsion.
4. Heat may increase the difference in density between the oil and the water, thus accelerating settling.

In general, at temperatures below 180°F (82°C), the addition of heat will increase the difference in density. Most light oils are treated below 180°F. For heavy crudes (below 20°API), heat may have a negative effect on difference in density.

In some cases, increased heat may cause the density of the water to be less than that of oil, as is shown in Figure 4.16. Adding heat changes the quality of the oil. The light ends are boiled off, and the remaining liquid has a lower API gravity and thus may have a lower value. Figures 4.17 and 4.18 illustrate typical gravity and volume losses for 33°API crude oil versus temperature. The molecules leaving the oil phase may be vented or sold. Heat can be added to the liquid by a direct heater, an indirect heater, or any type of heat exchanger.

A direct fired heater is one in which the fluid to be heated comes in direct contact with the immersion-type heating tube or element of the heater. These heaters are generally used when large amounts of heat input are required and to heat low-pressure noncorrosive liquids. These units are normally constructed so that the heating tube can be removed for cleaning, repair, or replacement.

An indirect fired heater is one in which the fluid passes through the pipe coils or tubes immersed in a bath of water, oil, salt, or other heat transfer medium that, in turn, is heated by an immersion-type heating tube similar to that used in the direct fired heater. Those heaters generally are used for high-corrosive fluids and are more expensive than direct fired heaters.

Heat exchangers are very useful where waste heat is recovered from an engine, turbine, or other process stream or where fired heaters are prohibited.

4.9.1 Treating Equipment

All devices that accelerate the separation of two phases when the natural retention time is too long for commercial application, usually over 10 min,

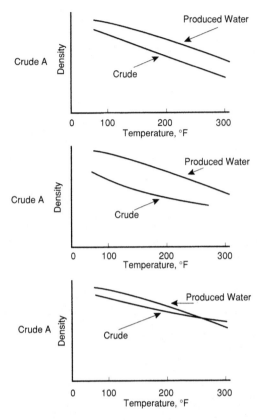

FIGURE 4.16   Relationship of specific gravity with temperature for these three crude oils [13].

are called *treaters*. Emulsion treaters use some combination of heat, electricity, chemicals, retention time, and coalescence to separate oil and water. Treaters are designed as either vertical or horizontal vessels. The vertical heater is shown in Figure 4.19. Flow enters the top of the treater into a gas separation section. This section can be small if the treater is located downstream of the separator. The liquids flow through a downcomer to the base of the treater, which serves as a free-water knockout; the bottom section can be small. If the total wellstream is to be treated, this section should be sized for sufficient retention time to allow the free water to settle out. This minimizes the amount of fuel gas needed to heat the liquid stream rising through the heating section. The oil and emulsion flows upward around the fire tubes to a coalescing section, where sufficient retention time is provided to allow the small water droplets to coalesce and to settle to the water section. Treated oil flows out the oil outlet. The oil level is maintained by pneumatic or lever-operated dump valves.

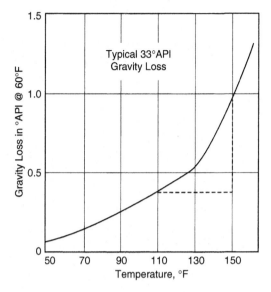

FIGURE 4.17 API gravity loss versus temperature for crude oil [13].

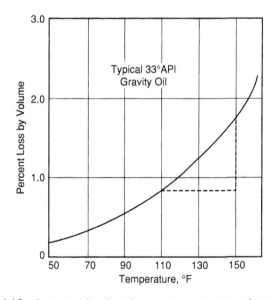

FIGURE 4.18 Percentage loss by volume versus temperature for crude oil [13].

It is necessary to prevent stream from being formed on the fire tubes. This can be done by employing the "40° rule." That means that the operating pressure is kept equal to the pressure of the saturated steam at a temperature equal to the operating temperature plus 40°F (4.5°C). The normal full-load

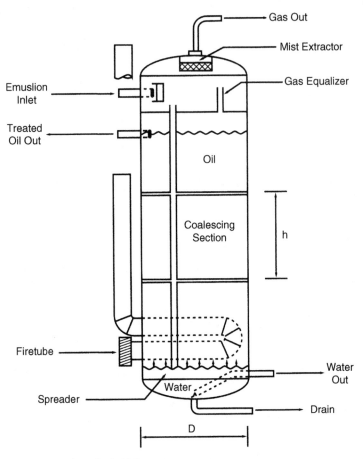

FIGURE 4.19   Vertical treater [15].

temperature difference between the fire tube wall and the surrounding oil is approximately 30°F, allowing 10°F for safety; the 40° rule will prevent flashing of steam on the wall of the heating tube.

A low-pressure vertical flow treater, of large diameter, is called a "gunbarrel" (Figure 4.20). Most gunbarrels are unheated, although it is possible to provide heat by heating the incoming stream external to the tank, installing heat coils in the tank, or circulating water to an external heater in a closed loop as shown in Figure 4.21. A heated gunbarrel emulsion treater is shown in Figure 4.22.

For higher flow, inlet horizontal treaters are normally preferred. A typical design of a horizontal treater is shown in Figure 4.23. Flow enters the front section of the treater where gas is flashed. The liquid flows downward to near the oil–water interface, where the liquid is water-washed and the free water is separated. Oil and emulsion rises past the fire tubes and flows into

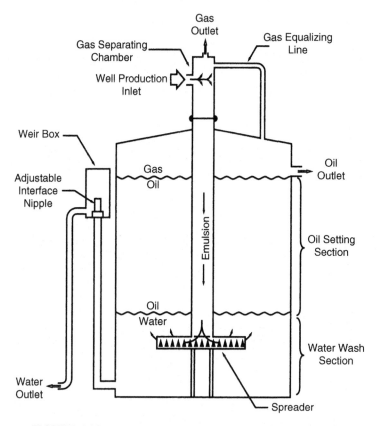

FIGURE 4.20 Low-pressure settling tank with internal flume [20].

an oil surge chamber. The oil–water interface in the inlet section of the vessel is controlled by an interface level controller, which operates a dump valve for the free water. The oil and emulsion flows through a spreader into the back or coalescing section of the vessel, which is fluid packed. The spreader distributes the flow evenly throughout the length of this section. Treated oil is collected at the top through a collection device used to maintain uniform vertical flow of the oil. Coalescing water droplets fall counter to the rising oil. The oil–water interface level is maintained by a level controller and dump valve for this section of the vessel. A level control in the oil surge chamber operates a dump valve on the oil outlet line regulating the flow of oil out the top of the vessel and maintaining a liquid-packed condition in the coalescing section. Gas pressure on the oil in the surge section allows the coalescing section to be liquid-packed. The inlet section must be sized to handle separation of the free water and heating of the oil. The coalescing section must be sized to provide adequate retention time for coalescence

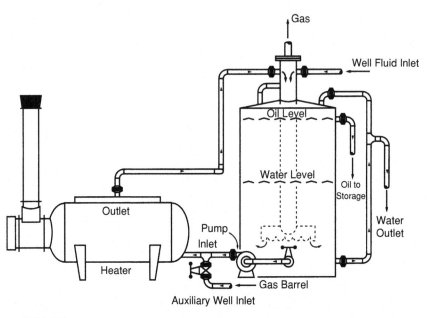

FIGURE 4.21   Heater and gunbarrel is forced circulation method of heating [20].

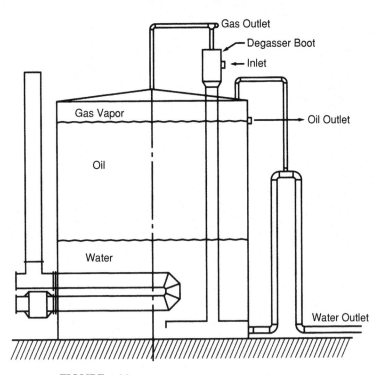

FIGURE 4.22   Heated gunbarrel emulsion treater [20].

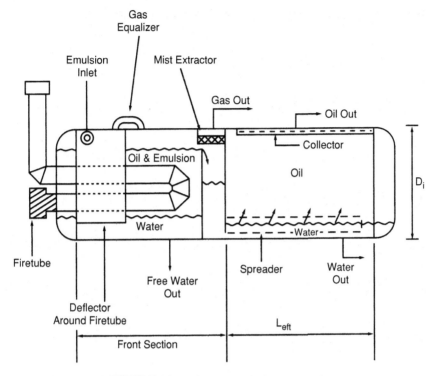

FIGURE 4.23   Horizontal heater-treater [20].

to occur and to allow the coalescing water droplets to settle downward counter to the upward flow of the oil.

4.9.2 Heat Input Requirements [21, 22]

The required heat input for an insulated vessel can be calculated as

$$Q_{th} = \dot{w}c\Delta T$$

where Q_{th} = heat required in Btu/hr
  $\dot{w}$ = weight flowrate in lb/hr
  c = specific heat constant at average temperature
    [Btu/(lb°F)] = 0.5 for oil and = 1.0 for water
  Δt = temperature increase, assuming a water weight
    of 350 lb/bbl

$$Q_{th} = \tfrac{350}{24} q_0 (SG_0)(0.5)\Delta T + \tfrac{350}{24} q_w (SG_w)(1.0)(\Delta T)$$
$$= 14.6\Delta T (0.5 q_0 SG_0 + q_w SG_w)$$

If heat loss is assumed to be 10% of the heat input, then

$$Q = \frac{Q_{th}}{0.9} = 16.2\Delta T(0.5q_0 SG_0 + q_w SG_w) \tag{4.10}$$

An alternative way is to employ the basic heat transfer equations, which are used in indirect heat sizing as follows:

$$Q = V_0 A T_m \quad \text{or} \quad A = \frac{Q}{V_0 T_m} \tag{4.11}$$

where Q_0, Q_w = heat required in Btu/hr
  A = total heat transfer area (coil area) in ft.^{2}
  T_m = log mean temperature difference in °F

For low-pressure oil, about 35°API and water liquid, streams, the heat required may be determined from the following equation:

$$Q = q_t[6.25 + 8.33(X)]\Delta T \tag{4.12}$$

where q_t = total liquid flowrate in bpd
  X = decimal water content in liquid
  ΔT = the difference between inlet and outlet
    temperature °F

The overall film or heat transfer coefficient for high-pressure gas streams may be found from Figure 4.24 using the gas flowrate and tube size selected. The overall film or heat transfer coefficient for water may be found from Figure 4.25 and the coefficient for oil from Figure 4.26, based on liquid flowrate and tube size. For liquid streams that are a mixture of oil and water, the overall coefficient may be averaged and calculated as

$$V_{0(mix)} = V_{0(oil)} + \left[V_{0(water)} - V_{0(oil)}\right] X \tag{4.13}$$

These film or heat-transfer coefficients are based on clean tubes; in other words, no allowance is made for any fouling factors. If any fouling is to be expected, excess coil area should be allowed in the heater selection.

$$T_m = \frac{GTD - LTD}{\ln\left(\dfrac{GTD}{LTD}\right)} \tag{4.14}$$

where T_m = log mean temperature difference in °F
  GTD = greater temperature difference
    = water bath temperature − inlet fluid temperature
  LTD = least temperature difference
    = water bath temperature − outlet fluidtemperature in °F

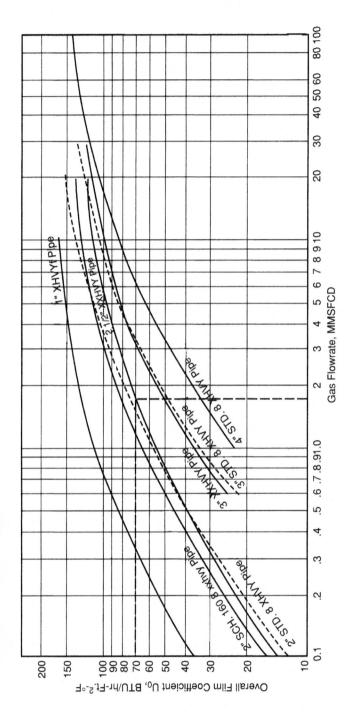

FIGURE 4.24   Overall film coefficient for natural gas in indirect heaters [22].

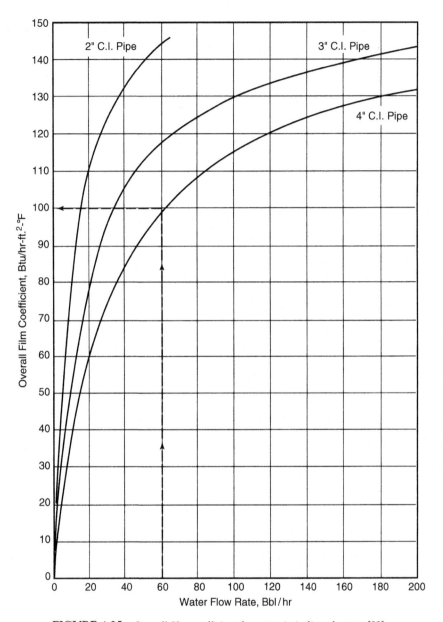

FIGURE 4.25   Overall film coefficient for water in indirect heaters [22].

A water bath temperature must be assumed for the calculations as mentioned before. Usually 180°F is the maximum designed temperature recommended for indirect water bath heaters.

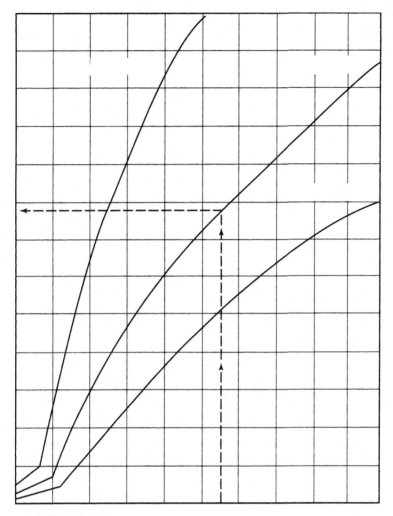

FIGURE 4.26 Overall film coefficient for oil in indirect heaters [22].

The coil area (A) is required for indirect heaters (Figure 4.27) [23] can be calculated from the basic heat transfer equation after all the above factors have been determined. An indirect heater then may be selected from the standard models listed in Tables 4.8 and 4.9 [22] based on the heat and the coil area required.

By selecting a heater with a larger heat capacity and coil area than that calculated, sufficient excess will be provided to allow for heat loss from the vessel and any fouling that may occur within the tubes, and will allow the heater to be operated at less than the maximum design water bath temperature.

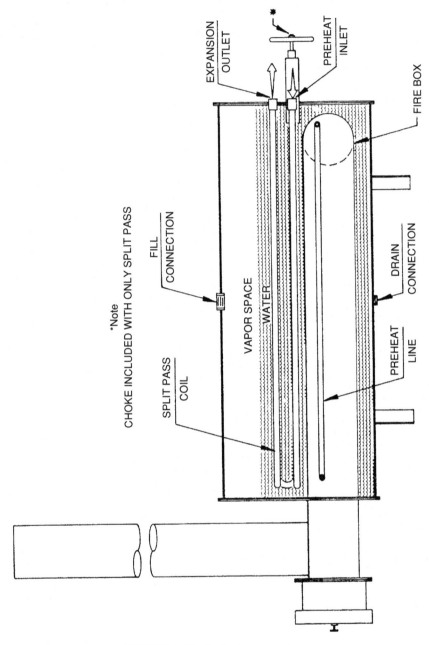

FIGURE 4.27   Indirect heater cutaway.

TABLE 4.8   Standard Indirect Heaters with Steel Coils [22]

Size Dia × Len.	Firebox Rating BTU/hr	No. of Tubes	Tube Size	Coil Area Sq ft	Split Pass Tubes Split	Split Areas	Equivalent Len. Of Pipe for Press. Drop, ft.
24″ × 3′	100,000	Spiral	1″ X	21.5	—	—	93.2
30″ × 6′	250,000	14	1″ X	26.2	8/6	15.0/11.2	96.4
30″ × 10′	500,000	8	2″ X or XX	28.9	4/4	14.5/14.4	68.3
36″ × 10′	750,000	8	2″ X or XX	48.8	4/4	24.4/24.4	100.3
		14	2″ X or XX	85.3	8/6	48.7/36.6	177.7
					10/4	60.9/24.4	
48″ × 10′	1,000,000	8	3″ X or XX	72.3	4/4	36.1/36.2	112.7
		18	2″ X or XX	109.7	12/6	73.1/36.6	229.3
		14	3″ X or XX	126.4	8/6	72.2/54.2	200.7
					10/4	90.3/36.1	
					12/2	108.3/18.1	
60″ × 12′	1,500,000	24	2″ X or XX	176.1	16/8	117.4/58.7	354.7
		20	3″ X or XX	217.1	12/8	130.3/86.8	328.7
					14/6	152.0/65.1	
		14	4″ X or XX	196.4	8/6	112.2/84.2	243.8
					10/4	140.3/56.1	
					12/2	168.3/28.1	
72″ × 12′	2,000,000	38	2″ X or XX	278.7	20/18	146.7/132.0	563.3
					28/10	205.4/73.3	
					30/8	220.0/58.7	
		30	3″ X or XX	325.6	22/8	238.8/86.8	495.4

(Continued)

TABLE 4.8   (Continued)

Size Dia x Len.	Firebox Rating BTU/hr	No. of Tubes	Tube Size	Coil Area Sq ft	Split Pass Tubes Split	Split Pass Split Areas	Equivalent Len. Of Pipe for Press. Drop. ft.
60″ × 20′	2,500,000	18	4″ X or XX	252.5	26/4	282.2/43.4	315.1
		24	2″ X or XX	295.5	12/6	168.3/84.2	546.7
		20	3″ X or XX	363.7	16/8	197.0/98.5	488.7
					12/8	218.2/145.5	
		14	4″ X or XX	328.4	14/6	254.6/109.1	355.3
					8/6	187.7/140.7	
					10/4	234.6/93.8	
					12/2	281.5/46.9	
60″ × 24′	3,000,000	24	2″ X or XX	355.2	16/8	236.8/118.4	642.7
		20	3″ X or XX	437.0	12/8	262.2/174.8	568.7
					14/6	305.9/131.1	
		14	4″ X or XX	394.3	8/6	225.3/169.0	411.8
					10/4	281.6/112.7	
					12/2	338.0/56.3	
72″ × 24′	4,000,000	30	3″ X or XX	655.4	22/8	480.6/174.8	855.4
					26/4	568.0/87.4	
		18	4″ X or XX	507.0	12/6	338.0/169.0	531.1

TABLE 4.9 Standard Indirect Heaters with Cast Iron Coils [22]

Size Dia × Len.	Rating BTU/hr	No. of Tubes	Tube Size	Coil Area sq ft (outside tubes surface)	Equivalent Len. of Pipe for Press. Drop, ft.
30″ × 10′	500,000	6	2″C.I.	33.6	65.4
		4	3″C.I.	33.3	47.7
36″ × 10′	750,000	14	2″C.I.	77.4	154.0
		8	3″C.I.	65.5	97.9
48″ × 10′	750,000	20	2″C.I.	110.2	220.5
48″ × 10′	1,000,000	18	2″C.I.	99.3	198.4
		12	3″C.I.	97.8	148.2
		18	3″C.I.	146.2	223.5
60″ × 12′	1,500,000	24	2″C.I.	162.0	312.8
		24	3″C.I.	238.5	346.9
		16	4″C.I.	209.6	250.5
72″ × 12′	2,000,000	38	2″C.I.	256.0	496.0
		38	3″C.I.	377.1	550.7
		20	4″C.I.	261.8	314.1

Example 4.6

For a given date, calculate the heater size such that

Oil flowrate $= 2,500\,\text{bpd}$
Water flowrate $= 1,400\,\text{bpd}$
Inlet temperature $= 60°\text{F}$
Outlet required temperature $= 100°\text{F}$
Coil size required $= 3\,\text{in.}$ cast iron

Heat required, from Equation 4.12, is

$$Q = q_t(6.25 + 8.33X)T$$

$$q_t = 2,500 + 1,400 = 3,900\,\text{bpd}$$

$$X = \frac{1,400}{3,900} = 0.36 = 36\%$$

$$\Delta T = 110 - 60 = 40°\text{F}$$

$$Q = 3,900[6.25 + 8.33(0.36)]40 = 1,442,813\,\text{Btu/hr}$$

Heat transfer coefficient, from Equation 4.13, is

$$V_{0(\text{mix})} = V_{0(\text{oil})} + \left[V_{0(\text{water})} - V_{0(\text{oil})}\right]X$$

Oil flowrate is $q_0 = \dfrac{2,500}{24} = 104.2\,\text{bphr}$

From Figure 4.26, $V_{0(\text{oil})} = 35.7\,\text{Btu/(hr\,ft.}^2\,°\text{F)}$.

Water flowrate is $q_w = \dfrac{1,400}{24} = 58.4\,\text{bphr}$

from Figure 4.25, $V_{0(water)} = 117.5\,Btu/(hr\,ft.^2\,°F)$.

$V_{0(mix)} = 35.7 + (117.5 - 35.7)0.36 = 65.15\,Btu/(hr\,ft.^2\,°F)$

Log mean temperature difference, from Equation 4.14, is

$$T_m = \frac{GTD - LTD}{\ln \dfrac{GTD}{LTD}}$$

$$GTD = 180 - 60 = 120°F \quad LTD = 180 - 100 = 80°F$$

$$T_m = \frac{40}{\ln \dfrac{120}{80}} = 98.7°F$$

Coil area, from Equation 4.11, is

$$A = \frac{Q}{V_0 T_m}$$

where

$$A = \frac{1,442,813}{117.5 \times 98.7} = 124.4\,ft.^2$$

From Table 4.9

> Heater size: 60 in. × 12 ft.
> Firebox capacity: 1,500,000 Btu/hr
> Coil data: 243-in. cast iron tubes
> Coil area: 238.5 ft.^{2}

Example 4.7

Size a horizontal treater for given data:

Oil gravity: 33°API, $SG_0 = 0.86$ at 60°F
Oil flowrate: 6,000 bpd
Inlet oil temperature: 100°F
$SG_w = 1.03$.

Assume that 80% of the cross-sectional area is effective, retention time is 15 min, and treating temperature is 120°F or 150°F. Then oil viscosity = $\mu_0 = 5.5$ cp at 100°F, 4 cp at 120°F, and 2.5 cp at 150°F.

1. Settling equation:

$$V_t = V_0 \text{ (terminal velocity of water = velocity of oil)}$$

Flow around settling oil drops in water or water drops in oil is laminar, and thus Stokes law governs:

$$V_t = 2,600 \frac{\gamma_w - \gamma_0}{\mu_0} D_p^2 \,(\text{ft./s})$$

$$\gamma_w, \gamma_0 (\text{lb/ft.}^3) \quad D_p(\text{ft.}) \quad \mu_0(\text{cp})$$

$$V_0 = \frac{Q}{A}$$

$$Q = 6.49 \times 10^{-5} Q_0$$

$$Q(\text{ft.}^3/\text{s}) \quad A(\text{ft.}^2) \quad Q_0(\text{bpd})$$

$A = D \times L_{eff}$ is the highest cross-sectional area.

$$V_0 = \frac{6.49 \times Q_0}{10^5 DL_{eff}} = 6.5 \times 10^{-5} \frac{Q_0}{DL_{eff}}$$

$$2,600 \frac{\gamma_w - \gamma_0}{\mu_0} D_p^2 = 6.5 \times 10^{-5} \frac{Q_0}{DL_{eff}}$$

$$DL_{eff} = 2.44 \times 10^{-8} \frac{Q_0 \mu_0}{D_p^2 (\gamma_w - \gamma_0)} \quad \text{where } D_p \text{ is in ft.}$$

or

$$DL_{eff} = 2,267 \frac{Q_0 \mu_0}{D_m^2 (\gamma_w - \gamma_0)} \quad \text{where } D_m \text{ is in } \mu m$$

The diameter of water droplets to be settled from the oil (μm) is a function of viscosity of the oil (cp) and, according to Arnold and Stewart [15], can be expressed as

$$d_m = 500(\mu_0)^{-0.675}$$

$$\text{at } 100°F \ d_m = 500(5.5)^{-0.675} = 158 \,\mu m$$

$$120°F \ d_m = 500(4)^{-0.675} = 196 \,\mu m$$

$$150°F \ d_m = 500(2.5)^{-0.675} = 270 \,\mu m$$

$$\gamma_m = 1.03 \times 62.4 = 64.3 \,\text{lb/ft.}^3$$

Assume the same for temperatures 60°F, 100°F, 120°F, and 150°F.

$$SG_0 \text{ at } 60° F = \frac{141.5}{API + 131.5} = \frac{141.5}{32.8 + 131.5} = 0.860$$

$$SG_0 \text{ at } 100° F = \frac{141.5}{(API - loss) + 131.5} = \frac{141.5}{32.8 + 131.5}$$

$$= 0.861$$

where "loss" from Figure 4.17 is $0.3 - 0.1 = 0.2$:

$$SG_0 \text{ at } 120°F = \frac{141.5}{32.7 + 131.5} = 0.862$$

$$SG_0 \text{ at } 150°F = \frac{141.5}{32.2 + 131.5} = 0.864$$

Because specific gravity of oil as a function of temperature changes in a small range, assume constant values for oil and water specific weights:

$$\gamma_w = 64.3 \, lb/ft.^3$$

$$\gamma_0 = 53.7 \, lb/ft.^3$$

$$\gamma_w - \rho_0 = 106 \, lb/ft.^3$$

	$T = 100°F$	$T = 120°F$	$T = 150°F$
$\gamma_w = 64.3 \, (lb/ft.^3)$	10.6	10.6	10.6
$\mu_0 \, (cp)$	5.5	4.0	2.5
$d_m \, (\mu_m)$	158.0	196.0	269.0

Calculate D versus L_{eff}

a. $DL_{eff} = 2{,}267 \times \dfrac{6{,}000 \times 4}{(196)^2 10.6} = 133.6$ if $T = 120°F$

b. $DL_{eff} = 2{,}267 \times \dfrac{6{,}000 \times 25}{(-296)^2 10.6} = 44.3$ if $T = 150°F$

D(ft.)	$L_{eff}(a)$	$L_{eff}(b)$
2.0	66.8	22.2
4.0	33.4	11.1
4.5	30.0	9.9
5.0	26.7	8.9
6.0	22.3	7.4
9.0	14.8	4.9
15.0	8.9	3.0
30.0	4.5	1.5

2. Calculate the retention time, such that

$$t = \text{Vol.}/Q \quad \text{Vol.} = (\pi D^2/4) L_{eff} \times 0.8 = 0.63 D^2 L_{eff}$$

$$Q = 6.49 \times 10^{-5} Q_0$$

$$t = \frac{0.63 D^2 L_{eff}}{6.49 \times 10^{-5} Q_0} = 9{,}681 \frac{D^2 L_{eff}}{Q_0}$$

$$D^2 L_{eff} = 0.000103 Q_0 t$$

$(t_r)_0$ is in min

$$t = 60(t_r)_0$$

$$D^2 L_{eff} = 0.00618 Q_0 (t_r)_0$$

If the retention time $(t_r)_0 = 15\,\text{min}$, then

$$D^2 L_{eff} = 0.00618 \times 6{,}000 \times 15 = 556.2$$

D(ft.)	2	4	4.5	5	6	9	15	30
L_{eff}(ft.)	139	24	27.5	22.2	15.5	6.9	2.5	0.6

If the retention time $(t_r)_0 = 20\,\text{min}$, then

$$D^2 L_{eff} = 0.00618 \times 6{,}000 \times 20 = 741.6$$

D(ft.)	2	4	5	6	9	15	30
L_{eff}(ft.)	185	46	29.7	20.6	9.2	3.3	0.8

Plot Figure 6.18, D versus L_{eff} gives the solution.
For settling temperature 150°F, the minimum effective length has to be 2.7 ft. if $(t_r)_0 = 20\,\text{min}$ and 3.4 ft. if $(t_r)_0 = 15\,\text{min}$. At lower settling temperature, L_{eff} should be 24 and 32 ft. for $(t_r)_0 = 20\,\text{min}$ and 15 min, respectively.
3. The heat required, from Equation 4.10, is

$$Q = 16.2 \Delta T \, (0.5 q_0 SG_0 + q_w SG_w)$$

or

$$Q_{th} = 14.6 \Delta T \, (0.5 q_0 SG_0 + q_w SG_w)$$

$$q_w = 0.1\% \text{ of oil flow rate}$$

$$\Delta T = 150 - 100 = 50°F \text{ and } 120 - 100 = 20°F$$

If $\Delta T = 50°F$, then

$$Q_{thl} = 14.6 \times 50 \times (0.5 \times 6,000 \times 0.861 + 600 \times 1.03)$$
$$= 2,336,730 \, Btu/hr$$

if $\Delta T = 20°F$, then

$$Q_{thII} = 14.6 \times 20 \times (0.5 \times 6,000 \times 0.861 + 600 \times 1.03)$$
$$= 934,692 \, Btu/hr$$

4. Conclusions are as follows. Treating temperature plays a more important role than retention time. For $T = 150°F$ and $L_{eff} = 7.3 \, ft.$, any diameter above retention time curves is correct. The diameter of the front section has to be the same as the coalescing section.

An economical solution would be a $6 \times 20 \, ft.$ for the coalescing section and a 2.5 Btu/hr firebox rating.

References

[1] Department of Interior—Minerals Management Service, 30 CFR Part 20, *Oil and Gas and Sulfur Operations in the Outer Continental Shelf*, 1995, http://www.mms .gov/federalregister/PDFs/docref.pdf.

[2] Texas Administrative Code Title 16, Part 1, Chapter 3, Rule §3.36, "*Oil, Gas or Geothermal Resource Operation in Hydrogen Sulfide Areas*," April, 1995, http://www.rrc.state.tx.us/ rules/16ch3.html.

[3] New Mexico Oil Conservation Division, Rulebook, Rule 118, NMAC Citation Number 19 NMZC 15.3.118, http://www.emnrd.state.nm.us/ocd/ocdrules/oil&gas/rulebook/ RULEBOOK.DOC.

[4] Title 49 CFR Part 191—Transportation of Natural and Other Gas by Pipeline; Annual Reports, Incident Reports, and Safety-Related Condition Reports, http:// www.access.gpo.gov/nara/cfr/waisidx/49cfr191.html.

[5] Title 49 CFR Part 192—Transportation of Natural and Other Gas by Pipeline: Minimum Federal Safety Standards, http://www.access.gpo.gov/nara/cfr/waisidx_02/ 49cfr192_02.html.

[6] Title 40 CFR Part 195—Transportation of Hazardous Liquids by Pipeline, http://www .access.gpo.gov/nara/ cfr/waisidx_02/49cfr195_02.html.

[7] ASME Pressure Vessel Code, http://www.asme.org/ bpvc/.

[8] National Board of Boiler and Pressure Vessel Inspectors, http://www.nationalboard. org/.

[9] API 510—Pressure Vessel Inspection Code: Maintenance, Inspection, Rating, Repair and Alteration, 8th Edition, June 1997.

[10] API Standards and Technical Publications, http://api-ep.api.org/filelibrary/ACF47. pdf.

[11] *Engineering Data Book*, Vols. 1 and 2, 10th Edition, GPSA, Tulsa, OK, 1987.

[12] Texas Natural Resource Conservation Commission–Standard Exemption 67, http:// www.tnrc.state.tx.us/permitting/airperm/nsr_permits/oldselist/889list/62–72.htm#67.

[13] e-Valves.com. E-Commerce Support for Valves, http:// www.evalves.org/

[14] Perry, R. H., and Green, D., *Perry's Chemical Engineers Handbook*, 6th Edition, McGraw-Hill, Inc., New York, 1984.

[15] Arnold, K., and Steward, M., *Surface Production Operation*, Gulf Publishing Co., Houston, Texas, 1986.

[16] Robertson, J. A., and Clayton, T. C., *Engineering Fluid Mechanics*, Houghton-Mifflin Co., Boston, 1985.

[17] API Specification 12J, 6th Edition, API, Washington, DC, June 1, 1988.

[18] Campbell, J. M., *Gas Conditioning and Processing*, Vols. 1 and 2, Norman, OK, 1984.

[19] Sivalls, C. R., *Oil and Gas Separation Design Manual*, Sivalls, Inc., Odessa, Texas.

[20] Bradley, B. H., *Petroleum Engineers Handbook*, Society of Petroleum Engineers, Richardson, Texas, 1987.

[21] API Specification 12K, 6th Edition, API, Washington, DC, June 1, 1988.

[22] Sivalls, C. R., Technical Bulletin 113, Sivalls, Inc., Odessa, Texas.

[23] Sivalls, C. R., *Indirect Heaters Design Manual*, Sivalls, Inc., Odessa, Texas.

Index

Notes: Page numbers followed by "f" refer to figures; page numbers followed by "t" refer to tables.

F

G

H

I

L

Printed in the United States
By Bookmasters